The Encyclopedia of

HOW IT'S MADE

Edited by Donald Clarke

A&W Publishers, Inc., New York

Pictures supplied by Aerofilms: 102T; AGFA-Gavaert: 62L; Albright & Wilson Ltd: 55L; Allard Graphic Arts: 3B, 58B, 76/7B, 90, 101B, 110TR, 135R, 172; Allard Graphic Arts/M E International Ltd: 125; Allied Breweries: 19T; Alphabet & Image: 188L; ASEA: 190/1T; Aspect Picture Library: 46CR&BR, 179B; ASPRO/Andrew De Lory: 12, 13, 14; Axminster Carpets/J Goldblatt: 35, 36, 37; Barmotts (London) Limited/Chris Barker: 22TL, TR, C, BL, CR&BR, 23TL&L; BASF UK Ltd: 7, 128L, 134BR; Borden (UK) Limited: 98/9; BPIF: 139T, 140/1B; Bridon: 161T; British Industrial Plastics Ltd: 129T&B; British Kinematograph Sound & TV Society: 108BR; British Ropes Ltd: 190B, 194B; British Steel Corporation: 28TL; Bryant & May: 104B, 105; James Burn Ltd/Andrew De Lory: 21TL, TR, B, CR&BR; Catomance Ltd/Varda Zisman: 52T&B, 53TL; Cerdep Ltd: 166T; Ciba-Giegy Ltd: 54; Clarke & Sherwell Ltd: 143TL; Clarks: 68L&R, 69T; Colorific: 57T, 77T, 88T, 186/7; Colour Workshops/David Kelly: 148/9, 151L&R; Combined Optical Industries Ltd: 95; Corah Ltd: 45; Courtaulds Ltd: 50L&TR, 51; Crown Paints & Wallcoverings Group: 113B, 115; Daily Mail: 154TL, TR, 155TL, TR&B; Daily Telegraph: 152L; Dartington Glass/M. Newton: 74TL, TR, CR&BR; De Beers Consolidated Mines: 64B, 65CL; De Lane Lea Ltd/Ian Duff: 111BL; Decca Photos/John Goldblatt: 157T&B, 158TL&TR, 159BL&BR; Deutches Museum: 3T; Dista Products: 119TR&BR; Dolby Labs: 110B; Dunlop: 4, 69B, 170TL&TR, 171T&B, 173, 174; English China Clays: 38TR, TL, 39BL&BR; Mary Evans Picture Library: 137R; Experimental Cartography Unit/Nerc: 100B, 102B; Fidor: 97; John R Freeman & Co: 99; Ford Motor Co: 116B; GKN: 167L&B, 168TL, TR&CR; Grove Park Studios: 108/9T; Guiness Breweries: 18; Sonia Haliday: 43B, 186, 188R; Hall's Barton Ropery Co Ltd: 163B; Claus Hansmann: 80L&R; Robert Harding Association: 163T; Richard Harrington: 56BL; Michael Holford Library: 42R; ICI Limited: 116T, 127B, 128R; International Wool Secretariat: 55R; Jackson Day Designs: 193; Eric Jewell Associates: 6, 127, 31; David Kelly: 150T(2); Frank Kennard: 198T; Keystone Press Agency Ltd: 114L&T; John Lang/Peter McQuinn: 17; Lea Valley Colour: 62TR, 63TL&TR; John Lobb Bootmaker/J. Goldblatt: 67; London College of Furniture: 71L; Nigel Luckhurst: 82; Wm Macquitty: 106/7B; Colin Maher: 184T, 186/7B; The Mansell Collection: 130, 133T, 153T; John Mathers/ Optical Surfaces Ltd: 93, 94T; McMullens/Brierley: 19B; Mike St Maur Sheil: 123, 124T, 126T; D Meredew Holdings Ltd: 71, 72L&C; Metal Box Company/Michael Newton: 29, 30; Mind & West Ltd: 70L; Monotype Corporation Ltd: 139C&B; Margaret Murray: 72R, 87; Nairn Floors Ltd: 95B, 96; Ordnance Survey: 100T; Osborne/Marks: 2BR, 2/3TR, 33, 34, 44, 65R, 66, 81, 103B, 104T, 118/9B, 131B, 136B, 140T, 144T, 146B, 160T, 162, 165B, 166B, 185, 192B; Parker Pen Co Ltd: 84; Payne & Co/Paul Brierley: 85; Perschke Service Organisation: 141R; Pfizer Inc: 8/9; Photo Syntax Films: 109B; Photri: 112, 127T, 133B, 156TL&B; Picturepoint: 11, 132/3B, 134TR, 177L, 184B; Pilkington Glass: 75L&R, 76TL&TR; John Pinchis: 60T; Pira: 135TL&BL, 142B, 143TL, 144C&B; Pittard Group: 88B, 89T&B; Popperfoto: 9TR, 10TL, 41TL, 121; Porvair Ltd: 90, 92; Price's Candles: 2BL, 24TL, TR, CR, BR, 25TL&TR, 26TL, CL&BR; Rainbox Colour/M. Newton: 62BR, 63BL&BR; Rank Film Labs: 109RC; Rank Zerox: 195, 196, 197, 198B; Alfred Reader & Co Ltd: 175; Max Redlich: 28CL; Reed Paper & Board (UK) Ltd: 117B, 118/9T, 120T; Research Consultants: 83, 86; Rockware Glass: 79; Roland Offset Maschinen/Fabrik Faber & Schleicherag: 147; Ronan Picture Library: 2TL, 15TR, 16B&T, 20BL, 59T, 103T, 131T; Royal Doulton: 40TL&BL, 40/1B, 41BL, 42L; Royal Mint: 60CR&BR, 61T&B; Samuelson Film Services Ltd: 106/7T; Schweppes: 28BL&CL; Science Museum: 47T; Shirley Institute Manchester: 53TR&BR; Slazenger Ltd: 176; SMPTE: 110B(1); Spectrum: 161B; St Bride Printing Library: 20BR; Stanley Tools: 180T&B, 181, 182, 183; Statsbibliothek Berlin: 23BR, 73, 132T; Steinmesse & Stollberg KG: 145; James Street Production: 108BL; Tate & Lyle Limited: 177R, 178T, 179T; Triplex Safety Glass Ltd: 78(2); UK AEA: 191B; United Spring & Steel Group Ltd: 194T; Viskase Limited: 164T; Visual Information: 48B; T Walls & Sons: 165T; John Watney: 134TL; Dr Harry Wignall: 47B, 48T, 49T&C; Yorkshire Post & Evening Post Ltd: 153C&B; Zefa: 32TL&BL, 46L&TR, 56/7BR, 57B, 65BL, 82, 117T; Zefa/Pictor: 136T, 152R.

First published in the United States of America in 1978 by
A & W Publishers, Inc.
95 Madison Avenue
New York, New York 10016
By arrangement with Marshall Cavendish Limited

Library of Congress Catalog Number: 78-58391
ISBN: 0-89479-035-8

Printed in the United States of America

INTRODUCTION

The many products that we use everyday – from the clothes we put on in the morning to the book we read at night – go through a process of development and manufacture which is interesting in itself. Even more important, the design decisions which are taken along the way affect the quality of our lives when the product finally comes to be used.

Matches, for example, are as safe today as they possibly could be, but early matches brought their convenience to the consumer only at the cost of an appalling industrial disease caused by the phosphorous. Similarly, in bookbinding the so-called 'perfect' binding, used until now only for work of indifferent quality, has been developed to the point where it will be possible to use this fast and inexpensive method for more permanent products.

The more we know about manufacturing methods, the more informed we become as consumers and the less inclined we will be to take goods and the information we are given about them at face value. The information in this volume will show that it is possible to make one's own decision about getting value for money, by simply knowing more about *How It's Made.*

CONTENTS

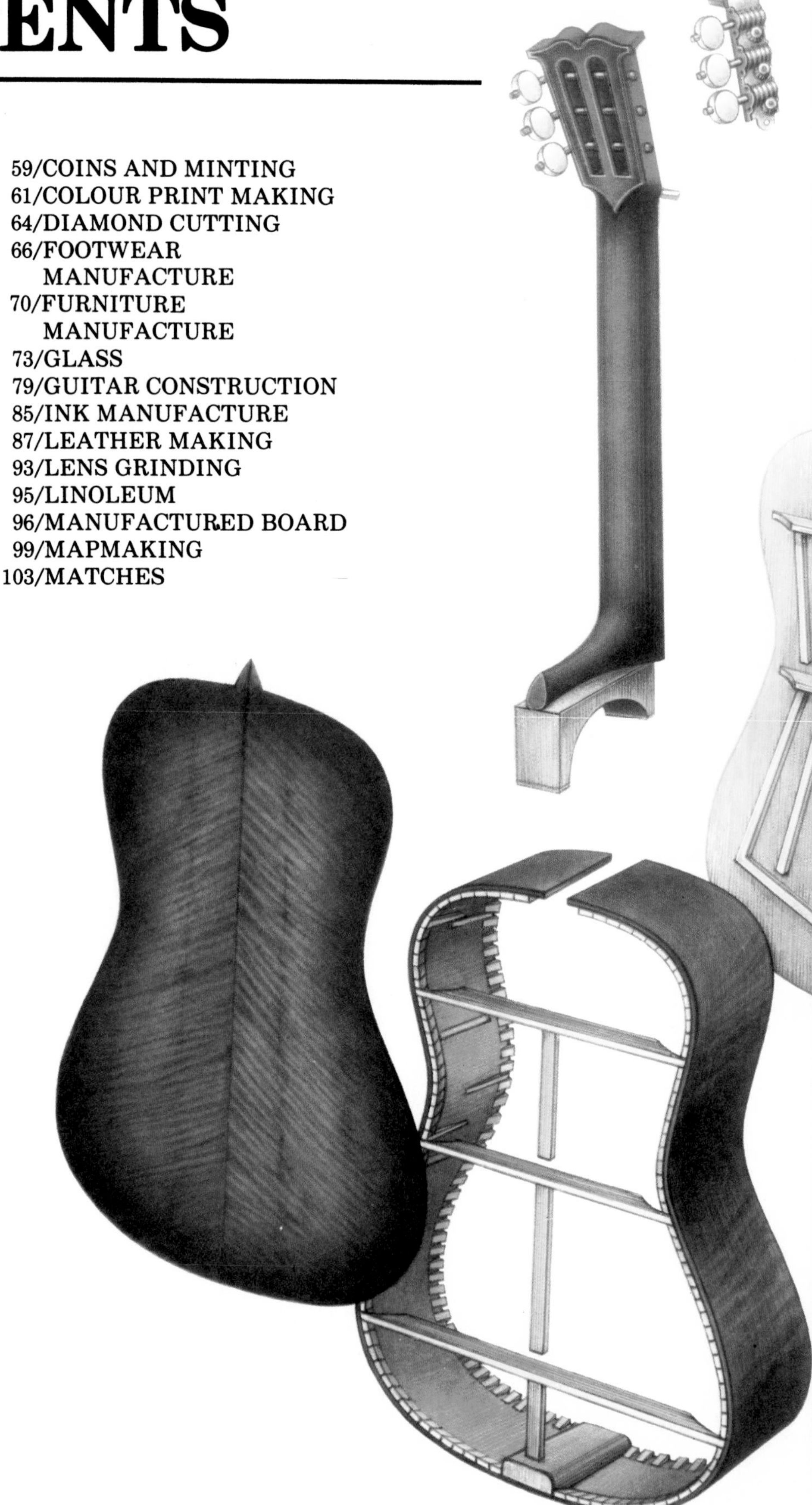

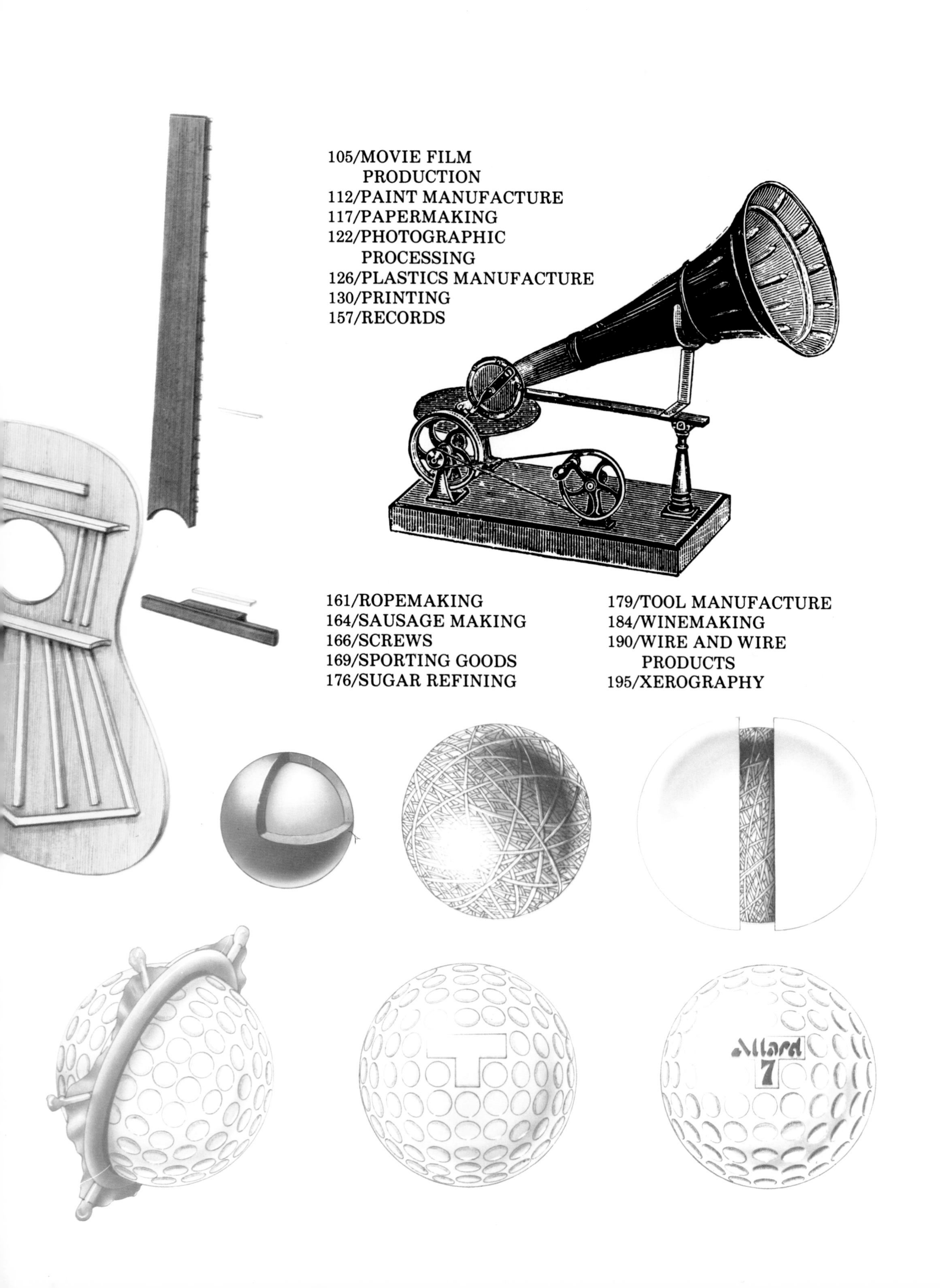
Allard
7

ACID manufacture

The quality that distinguishes an acid from other substances is that its molecules contain hydrogen atoms which partly split away from the rest of the molecule when the acid is dissolved in water. This causes the hydrogen atoms to become electrically charged ions with a strong tendency to react with other substances—hence the corrosiveness of many acids. The molecules of 'strong' acids have a great tendency to split.

The main acids manufactured and used in industry are sulphuric, nitric and hydrochloric acid—all of which are strong acids—and acetic acid, a relatively weak *organic* acid, that is, one with a chemical formula related to the complex carbon compounds found in living things.

SULPHURIC ACID This is a clear, oily liquid with the chemical formula H_2SO_4. It dominates the market for acids.

It can be manufactured directly from sulphur, or from anhydrite (calcium sulphate), a common mineral that is also used for making cement. Other sources include the sulphur-containing by-products of other industrial processes.

The original and traditional method of making sulphuric acid is by the chamber or tower process, so called because the main reaction takes place in a lead-lined chamber, and other parts of the process in towers. The acid is manufactured by burning sulphur to give sulphur dioxide and reacting this with air and steam in the presence of oxides of nitrogen, which act as catalysts. (A catalyst is a substance which assists or causes a reaction without itself being changed.) The reaction is complex, on account of the presence of the catalyst. The chamber process yields acid of a rather low strength and purity, and its use has dwindled until now only about 2% of sulphuric acid is made by it. It has been supplanted by the more sophisticated *contact* process, which gives very pure acid of any strength. It can even produce the acid in a 'super-charged' form called oleum, or fuming sulphuric acid, which has the chemical formula $H_2S_2O_7$. This intensely reactive and highly dangerous substance turns into ordinary H_2SO_4 when added to water; if, on the other hand, the water is added to the acid, the reaction boils the water violently, spraying water and acid in all directions.

Opposite page: the converter of a modern sulphuric acid plant using the contact process. It consumes roughly 500 tons of sulphur a day.

Below: solid sulphur is melted and burned. It combines with oxygen to make sulphur dioxide (SO_2). The heat from the process is used to melt more sulphur.
The SO_2 is then converted to sulphur trioxide (SO_3) and mixed with ready-made sulphuric acid (H_2SO_4) to make oleum ($H_2S_2O_7$), a super-concentrated form which is then diluted.

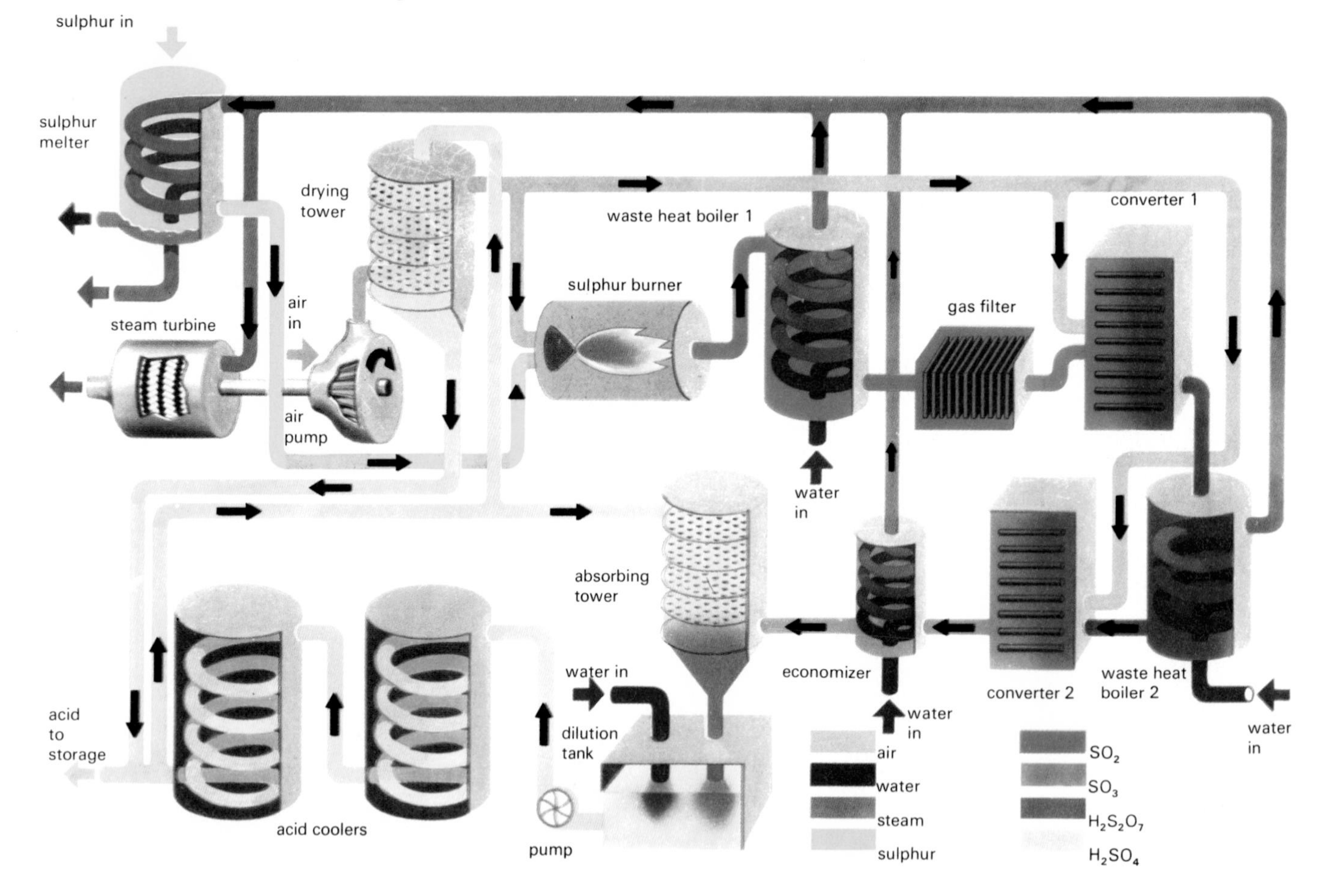

The outline of the contact process is shown in the diagram. In brief, sulphur or sulphur-containing material is burned with dry air to produce sulphur dioxide (the air can be conveniently dried by using some of the acid, which readily absorbs water). The sulphur dioxide is filtered, then passed to a converter, where more air is added in the presence of a catalyst (platinum or vanadium pentoxide) to convert it to sulphur trioxide.

This could now be added to water to make sulphuric acid, but the reaction is rather violent, so in practice it is added to the acid itself to make oleum. This can then be diluted with water.

Many stages of the process produce intense heat; this is controlled, and also used, in waste heat boilers. Water is pumped past the hot chemicals in a closed coil. The heat turns it to steam, which is then used in other parts of the process.

Of the sulphuric acid produced, about one third goes to make fertilizers. Other important uses are in the production of paints, pigments, fibres, detergents and plastics. It is also used for pickling (cleaning) steel, making dyestuffs and related products, and manufacturing other acids.

NITRIC ACID This acid has the formula HNO_3. It is a colourless, fuming liquid when in a pure state, but it is unstable and soon acquires a yellow or red colour when exposed to the air. This is caused by the presence of the gas nitrogen dioxide, which forms as the acid is decomposed by light or by high temperatures. The fumes of nitrogen dioxide are extremely poisonous, and the acid itself is one of the most corrosive known. It cannot be stored in a bottle with a cork or rubber stopper, since it attacks both these materials. It has to be transported in stainless steel or aluminium containers.

Nitric acid is produced in the laboratory (and was once produced in industry) by treating Chile saltpetre (sodium nitrate) with sulphuric acid. The modern industrial technique is to make the acid from ammonia (itself prepared by extracting nitrogen from the atmosphere), which is treated with air in the presence of a platinum-rhodium catalyst to produce nitric oxide. Further air is then admitted to the 'converter' vessel in which the reaction takes place. This turns the nitric oxide to nitrogen dioxide. Finally, the nitrogen dioxide is dissolved in water. With the help of more atmospheric oxygen, it forms nitric acid.

Nitric acid is used for making fertilizers, explosives, dyes and drugs, and also for etching, since it attacks almost all metals. A mixture of one part of nitric acid to three of hydrochloric acid, called *aqua regia*, will even dissolve gold.

HYDROCHLORIC ACID The formula of this acid is HCl. In the pure state it is a gas, but is always used and sold as a solution in water. It is extremely corrosive in either state, and is transported in glass or rubber-lined tanks.

Hydrochloric acid is most commonly manufactured by the electrolysis of brine (salt water). The reaction also produces caustic soda (sodium hydroxide).

Hydrochloric acid is used for pickling steel before it is galvanized (zinc-plated), for decomposing bones to make gelatine, in the manufacture of dyes and rayon, refining oils, fats and waxes, tanning leather and purifying silica.

ACETIC ACID This acid has a more complex structure than the other three mentioned above. Its chemical formula is conventionally written CH_3COOH, which describes its molecular structure to a certain extent as well as its content. It is prepared in various ways, the most important process using naphtha, a cheap and readily obtainable by-product of the petroleum industry.

Acetic acid is the principal ingredient of vinegar, giving it its sour taste (nearly all acids taste sour, but most are poisonous). It is used to make cheap synthetic vinegar, as used by the processed food industry. But this is only a tiny proportion of its main usefulness.

Uses include making cellulose acetate for synthetic fibres, plastics and packaging, vinyl acetate for emulsion paint and adhesives, acetate ester solvents for paint and plastics, synthetic fibres, and pharmaceuticals.

ANTIBIOTICS

Antibiotics are a useful group of chemical substances produced by certain types of micro-organisms. In low concentrations, they can kill or stop the growth of other micro-organisms which cause diseases in man, in animals and in plants.

True antibiotics are produced by moulds, bacteria, or actinomycetes (organisms intermediate between bacteria and moulds). Others may be chemically synthesized, or modified from the molecule of a naturally occurring antibiotic. Technically, they are then no longer true antibiotics but are just as effective in fighting diseases. Over 2000 antibiotics have been identified or synthesized, and of these, about 60 have been produced commercially.

The majority of the antibioitics are used in human and veterinary medicine as antibacterial agents; others are used to treat fungal infections, such as ringworm, in man and in animals, or fungal diseases affecting crops; some are effective against protozoal infections such as amoebic dysentry; a few are used in the treatment of some rare types of cancer.

Antibiotic research stems from the accidental discovery by Sir Alexander Fleming, in 1928, of a colony of *Penicillium* mould contaminating a laboratory culture of staphylococci, the bacteria causing boils and blood-poisoning. Substances produced by the mould colony had diffused out and killed the surrounding staphylococci. Tests showed that the mould could also kill or inhibit the growth of a number of other kinds of disease causing bacteria, but for several years researchers were unable to isolate the active agent, which was highly unstable. When penicillin was eventually isolated in very small quantities, clinical trials demonstrated its potential medical value, and the advent of World War II provided the impetus for large scale commercial development.

PRODUCTION TECHNIQUES British manufacturers commenced production using the only available method, which was basically a laboratory technique. The *Penicillium* mould was grown on the surface of a

nutrient liquid (see below) in thousands of glass flasks. Once the mould colony had spread to cover the surface of the nutrient, it was filtered off, and the crude penicillin, which had diffused from the mould into the liquid, extracted from the liquid. A very low yield was obtained, and there were additional problems with contamination of the culture by other micro-organisms.

Meanwhile manufacturers in the United States concentrated on improved methods, and perfected the technique of deep fermentation now used for the production of most antibiotics. Deep fermentation was based on the discovery of a species of *Penicillium* which would grow submerged in the nutrient liquid, rather than as a thin skin on the surface, thus producing a higher yield of penicillin from a given volume of liquid. Further strains and artificial mutations of *Penicillium* were produced, with improved yields, and finally, the discovery of a new liquid increased the yield ten-fold.

Above: Professor Alexander Fleming, who discovered penicillin in 1928.

Left: a symmetrical colony of green mould called Penicillium chrysogenum. *A mutant form of this has become very important because it is used to produce almost all commercial penicillin.*

In modern production, master cultures of selected strains of *Penicillium* are stored in controlled conditions, to ensure uniformity of subsequent production. Small sub-cultures are transferred to culture flasks, allowed to develop on a suitable liquid medium, then transferred to a larger vessel. Once again, the culture develops and is transferred to progressively larger fermenters until it reaches the final stage, where the fermenting tank, which is usually made of stainless steel, may be of 30,000 gallons (136,000 litres) capacity.

The mould grows submerged in a sterilized nutrient broth based on corn-steep liquor (a by-product of the starch manufacturing industry), sugar, salts, and

other carefully controlled ingredients which are used to modify the penicillin molecule. The final fermentation is complete after 1 to 2 weeks, when the contents of the tank are drawn off and filtered to remove the *Penicillium* mould. The remaining liquid, containing the penicillin, is chemically purified and concentrated. A final chemical process precipitates penicillin out in a fine crystalline form, which is filtered out, washed with a solvent to remove impur-

Above: a thirty-year-old penicillin culture flask, containing the growing mould. In the final stage, the mould ferments in large tanks.

Above right: two antibiotic crystallization units in operation with the pipe connection panel. First dissolved in preparation tanks, then sterilized by filtration, the antibiotics are conveyed to the crystallizers through the pipe panel, which can feed to either unit from any of several tanks.

Right: in the final stage, the crystals are washed and then dried in a vacuum. The standing man is checking mixture flow; the other is taking a sample of waste to make certain that no antibiotic is lost.

ities, dried and stored.

Each batch is rigorously tested to ensure its activity and purity, and the entire manufacturing process takes place under sterile conditions. By varying the constituents of the nutrient broth, chemically different forms of penicillin can be produced, and the penicillin molecule can be further modified by chemical treatment after purification to produce semi-synthetic penicillins which do not occur in nature. More extensive modifications to the basic molecule can produce entirely new antibiotics, such as ampicillin and cephalexin. Other types of antibiotic are manufactured using similar fermentation processes.

ANTIBIOTIC APPLICATIONS Penicillin and related antibiotics are extremely active against many different types of bacteria; other antibiotics have activity against more restricted groups of micro-organisms. Broad-spectrum antibiotics such as the tetracyclines and chloramphenicol, which is used for typhoid, can also be used for a wide variety of bacterial infections, and a similar effect is sometimes obtained by combining antibiotics with narrower ranges of activity.

Allergic reactions and side effects of varying severity may be experienced in a proportion of patients treated with antibiotics. In addition, antibiotic-resistant strains of bacteria have emerged, and diseases which formerly responded rapidly to antibiotic treatment sometimes cause problems, particularly in hospitals. For these reasons, the antibiotic activity of a wide range of micro-organisms is continuously monitored, in an attempt to discover broad-spectrum antibiotics with a minimum of undesirable side effects, and which it is hoped will prevent the emergence of resistant strains.

One result of such a screening programme was cephaloridine, derived from an antibiotic produced by *Cephalosporium*, a mould collected from the Mediterranean in 1945. In laboratory studies, 40,000 mutated strains of the mould were tested until one was selected, yielding sufficient quantities of the natural antibiotic substance, which was then chemically modified to produce cephaloridine. Further modification to the cephaloridine molecule produced cephalixin, which can also be prepared by modifying the penicillin molecule.

In addition to the above activities, some antibiotics have a marked growth promoting effect on livestock, and are often incorporated in small amounts into animal feeds for this purpose. Because of fears that antibiotic-resistant strains of bacteria might develop in treated livestock, and subsequently be transferred to humans, only a limited range of these antibiotics, none of which are used in human medicine, is permitted in animal feeds in many countries.

In Japan and the USA, much research has been carried out on antibiotics for horticultural purposes, mainly to combat fungal diseases of crops, such as rusts and smuts. Large scale production allows 10,000 tons of blasticidin-S to be used annually in Japan as a treatment for 'rice blast', a serious fungal disease of rice paddies.

ASPIRIN manufacture

Aspirin is the most widely used of all analgesics, or painkillers, and is used for the relief of mild pain and to reduce fevers. Together with some related compounds, aspirin is also used to reduce painful swelling and inflammation, particularly in disorders such as arthritis. The word 'aspirin' is actually a trade name for acetylsalicylic acid. This is prepared from two organic compounds, salicylic acid and acetic anhydride.

Inflammation of the stomach wall and gastric bleeding commonly occur in persons taking regular doses of aspirin. To alleviate this, aspirin may be supplied in a microfine or very finely divided form, or as a soluble tablets or powder, which are absorbed rapidly from the stomach. Aluminium acetylsalicylate is said to produce less gastric disturbance, as is salicylamide, which also has a sedative effect.

Phenacetin, another analgesic, is often combined with aspirin, but its consumption has declined through fears that prolonged usage may cause kidney damage. In the body, phenacetin is metabolised into paracetamol, which gives phenacetin its pain-relieving properties. Paracetamol itself is now widely used as a painkiller, since it lacks the anti-inflammatory properties of aspirin, does not cause gastric bleeding or digestive disturbances, and is much less toxic to the kidneys than phenacetin.

Codeine is another common constituent in Britain of compound analgesic tablets. It is a derivative of morphine, but is not known to be habit-forming. Codeine is used in very small quantities, even in 'Codeine'

Above: microphotograph of salicylic acid crystals.

tablets, which contain mostly aspirin. Caffeine is also frequently included in compound tablets, being used for its mild stimulant effects.

TABLET MANUFACTURE Aspirin and many of the analgesics described above can be taken as a powder, dissolved in water. Most analgesics, however, are supplied as tablets, allowing precise regulation of the dose, and minimising unpleasant taste.

The basic manufacturing technique is to compress the ingredients into a tablet, together with various excipients, or medically harmless additives which assist the process. The initial process is the manufacture of granules, which when compressed will adhere together firmly, but allow the tablet to disintegrate rapidly when swallowed. Aspirin tablets are usually manufactured by a process of double compression, or 'slugging'. Powdered aspirin and starch are pressed into extremely hard pellets, which are broken up to produce granules. This is a comparatively simple process, but granules for compound tablets are usually prepared by the more complex 'wet granulation' technique.

In wet granulation, after the constituents are mixed a binder such as acacia gum or tragacanth is added, serving to hold the powder together. The mixture is moistened, and the resulting doughy mass rubbed through sieves to produce granules. These are dried in a fluidized bed drier, where granules are 'floated' on an upward current of warm air until they have dried to the required moisture content. The entire sequence can be performed on a single machine.

Before final compression of granules produced by slugging or wet granulation, disintegrants are usually added, such as starch or alginates. On contact with liquid in the stomach, these will swell and cause rapid disintegration of the tablet. A lubricant material may also be added to prevent adhesion of the granules or finished tablets to the compression machine.

Some modern high speed rotary tablet presses can produce over half a million tablets an hour. Granules are fed from hoppers to the head of the machine, where metered amounts are dispensed into a die. Shaped upper and lower punches move together within the die, compacting the granules into a tablet. This is ejected into a chute as the head of the machine revolves to bring another set of punches into position, and the process is repeated continuously.

Some types of analgesics are supplied with a coating designed to dissolve only after the tablet has left the stomach and passed to the intestine. This relatively impervious layer is applied by placing tablets and coating material in a huge rotating drum, where they are rolled together until a sufficient layer of the coating builds up on the tablets. Multi-layer tablets are also available, designed to liberate the constituents at a varying rate, or at a specific point in the digestive system.

Effervescent tablets are another variant, to be dissolved in water before consumption. Compounds such as citric acid and sodium bicarbonate are incorporated, which in contact with water will react together, causing the tablet to froth and disintegrate quickly.

Above: the raw materials being fed into the plant. On the right are the vats containing powdered starch and salicylic acid, which go into the mixers in the centre. The semi-automatic control console is on the left.

Above right: the first step in aspirin production. The roller granulator takes the powder from the mill and forms it into a semi-solid.

Right: tablets coming from the feed frame of the tabletting machine.

Counterclockwise from left: finished tablets; the Auto Analyser, which measures the amount of aspirin in a tablet; a tablet disintegration test: the tablets are washed until they fall through a sieve; tablets fed into the hopper of a machine which arranges them in rows for packaging; the Monsano hardness tester, which measures the force needed to break the tablet; the pneumatic device for placing a tablet on electronic weighing equipment.

Above left: arrays of tablets passing an optical checking device. If there are any tablets missing in an array, the whole array is not packaged.

Above right: the tablets after the foil has been applied and before the tapes are cut.
The other two pictures show the finished foil-packs of tablets.

BEER and brewing

Beer has been made in various forms for at least six thousand years, and is known to have been made by the Babylonians and ancient Egyptians. In early times, brewing was a cottage craft carried out at the same time as baking, since the initial processes of brewing were then very similar to those of bread making. Primitive types of beer were produced by steeping partly cooked bread in water and allowing it to ferment. The bread was made from a mixture of crushed barley which had begun to germinate (sprout), and yeast.

By the 14th century AD, brewing had developed into a separate trade with its own specialized skills. To a considerable extent this was stimulated, over a period of three centuries or so, by the brewing activities of the monasteries. In mediaeval times the monks were the main producers of beer, serving not only their own needs but also those of the local people. Brewing was also carried out at home, mostly by the women.

As a trade, brewing then expanded steadily for some five hundred years, notably in Europe and, in the 18th century, especially in North America. By the middle of the 19th century many thousands of breweries were in operation in the western world. Since then the production of beer has become a major industry and modern breweries are large and complex. World production of beer in 1973 was about 15 times as great as in 1873; in the same period the number of breweries fell tenfold, so on average the 1973 brewery produced 150 times as much beer as its counterpart 100 years before. This very considerable expansion is also well illustrated in brewing technology: the first yeast separator, used commercially in brewing around 1898, had a throughput of 1 cubic metre (35 cu ft) per hour; the capacity of the 1973 version was 200 cubic metres (7073 cu ft) per hour.

TYPES OF BEER Beers can be divided into two main groups: lager or bottom fermented beers, and the top fermented British type beers. Some varieties of lager are known by the names of the places in which they were first brewed, such as Pilsener (from Pilsen, in Czechoslovakia) or Dortmunder (from Dortmund in Germany). Most of them are pale in colour, well aerated, and with a less pronounced hop flavour than British beers; bottom fermentation is also used to produce darker, more full bodied beers. Apart from the British Isles, most of the beer brewed throughout the world is of the lager type, generally with an alcohol content of between three and five per cent by weight.

Although an increasing amount of lager is being brewed in Britain, the top fermented ales (beers) and stouts (a type of dark brown beer) are still the most

Right: a mash tun in a London brewery, 1823. Behind the tun a man stokes a fire under a brewing copper.

Below: an 18th century London brewery. The mash tuns are in the foreground; they are stirred with wooden paddles. The brewing coppers are at the rear.

popular. Until the 17th century the name 'ale' referred to a drink brewed from malt, yeast, and water, whereas beer was made by the addition of hops during the brewing. Beer was brought to England from Europe in the 15th century, and by the early 18th century it had replaced ale as the main English brew. Although ale as such is no longer made, the name is still applied to any beer which is not a stout or a lager, for example Bitter Ale, Pale Ale, Brown Ale. The alcohol content of ale is usually between 2.5% and 6.5% by weight. Stout, which is brewed with roasted malt and often a high percentage of hops, has a strong, rich flavour. It is very dark, often black in colour, containing up to 5.5% of alcohol. Porter was a dark beer of the 18th and 19th centuries, full bodied but milder than stout.

THE BREWING PROCESS The raw materials used in brewing have a major influence on the type and quality of the beer produced. In theory, beer can be made by fermenting any cereal, or other source of starch such as potatoes, in water. In practice, barley is the most widely used cereal. Other cereals are used as additives to the main barley mash to reduce costs, and sometimes to produce a desired flavour. The main additives are rice, corn, tapioca, soyabean meal, unmalted barley, and various sugars.

The initial brewing operation is to make a liquid mixture from barley, water and hops which is known as the wort. Barley cannot be used, however, until it has been malted, a process not usually carried out at the brewery itself but at plants known as maltings.

Above: the mash tuns at the St James' Gate Brewery, Dublin, in 1890. It is far more mechanized and cleaner than the brewery of 100 years before.

Below: a 19th century malting. The grain was steeped for 40 hours in the tank at the far end, then heaped on the floor for 96 hours. Then it was spread across the floor and turned two or three times a day for 12 or 14 days.

Opposite page: a flow chart showing the brewing process. Used grains are used as cattle feed; spent hops may be used as fertilizer.

The malting of barley involves germinating it under controlled conditions to produce natural substances called enzymes, which act as catalysts in various chemical reactions vital to brewing.

To soften the barley and promote its germination, it is soaked in water at 13° to 16°C (55° to 60°F) for between 48 and 72 hours, depending on the type of grain used. After soaking, the barley is put into large drums or boxes, and moist air is blown through it for 7 to 11 days to encourage germination. It is then dried in a kiln until its moisture content is down to 1.5% to 2%. The rootlets which have grown during germination drop off and are used as animal food. The barley is now known as malt, and contains enzymes.

At the brewery the malt is crushed and made into a mash with water and the additives. Mashing sets the enzyme process in operation and brings out the soluble materials such as starch and sugar from the malt. Insoluble material such as protein is made soluble by action of enzymes, which also convert the malt starch into maltose sugar; the amount of maltose produced determines the alcohol content of the beer. The mashing operation must be very carefully controlled so that all the physical and chemical processes and enzyme reactions are coordinated to produce precisely the type and quality of wort, and therefore beer, which is required. The mashing process for lager beer is different to that for the British beers.

INFUSION MASHING The mashing for British beers is known as infusion mashing. It is carried out in large insulated tanks called tuns, which are generally heated by steam. The consistency of the mash is important, so the tuns are often fitted with mechanical agitators. Precise temperature control is vital, since a deviation of only a few degrees can produce a totally different type of wort from that required.

Once the warm mash has reached the point where starch conversion (to maltose sugar) is complete, the temperature is raised to about 75°C (167°F) for a short time. This operation, known as 'mashing off', is carried out to inactivate the enzymes, most of which stop working at this temperature. The mash is then allowed to stand for 30 minutes, to allow the insoluble grain husks to settle out. The husks form a layer on the false bottom inside the tun and act as a filter. The liquid wort is run through until it becomes clear; the 'spent' or used grains are washed ('sparged') with hot water to ensure that all the soluble matter passes through the false bottom of the mash tun into a receiving vessel.

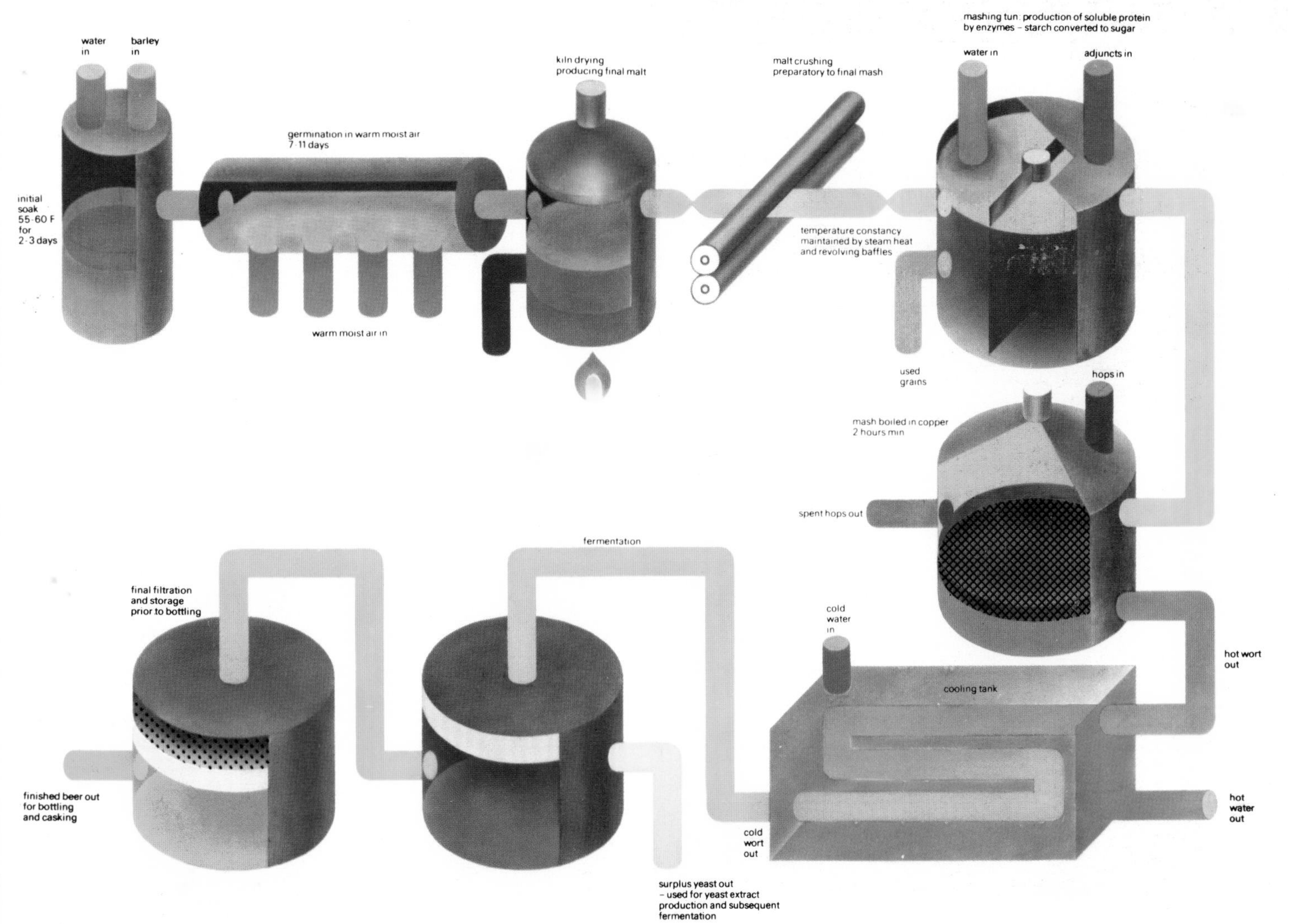

DECOCTION MASHING The malted barley used in lagers is not germinated for as long as that used in British beers, and so it needs to be more finely mashed. The mashing is done in stages: a preliminary mash at 37°C (100°F), followed by subsequent mashes at 50°C (122°F), 65°C (149°F) and 75°C (168°F); or the quick mashing system used in the USA with two mashes at 65°C (149°F) and 78°C (172°F).

BOILING After mashing, the wort and sparge water are transferred to a large copper vessel known as a brewing kettle, and boiled vigorously with hops or hop extracts (which are sometimes added progressively), for at least two hours. This operation does several things: it sterilizes the wort and reduces its bulk by evaporation of the water; it draws out the full bitter flavour of the hops and helps precipitate any unwanted protein left in the wort; and it ensures that if any enzymes have survived the 'mashing off' operation they are now made completely inactive, preventing spoilage of beer in cask or bottle by further reactions.

After boiling, the wort is discharged through a filter bed made from spent hops, and then cooled, usually by heat exchangers, and aerated, which helps fermentation later on.

FERMENTATION When the wort is at the optimum temperature for starting fermentation the yeast is added. Yeasts are microscopic organisms related to fungi, and there are thousands of different species. There are many strains of the brewer's yeast but they can all be placed into one of two groups: they either rise to the surface or sink to the bottom during fermentation, thus giving top fermented or bottom fermented beers. The particular temperature chosen depends on the quality and strength of the beer, and as yeast is a form of plant the fermenting temperature is also varied at different times of the year. Weaker beers require higher temperatures than stronger beers. The action of the yeast on the wort is extremely complex, producing alcohol and carbon dioxide as the principal products, and many other substances such as acids, esters and glycerine, all of which affect the final flavour and aroma of the beer.

For bottom fermented beers the yeast is added at a temperature of 6° to 10°C (43° to 50°F), and fermentation takes about eight days, after which the 'green' beer is put into storage tanks for up to three months (the name *lager* comes from the German word for 'storage'). The lager is stored at 0°C (32°F), and a secondary fermentation occurs which clears the beer and improves the flavour.

British beers begin fermentation at about 15°C (60°F), and during the process the temperature increases to about 21°C (70°F). The fermentation takes five to seven days, followed by a low temperature maturation period, perhaps three weeks.

Whichever process is used, the yeast layers are separated off and may be used in subsequent brews. Fermentation produces more yeast than can be used in this way, and the surplus is used for animal feeds and yeast extract manufacture. The beer may be given a very fine filtration in order to 'polish' it,

Opposite page: three pictures from the Guinness brewery at St. James' Gate, Dublin. From top: the mash tuns, the coppers, and pumping the hot wort into a wort cooler.

Above: fermentation vessels producing top-fermented English beer. The yeast forms a thick, creamy layer on top.

Below: filling metal kegs with beer, a process known as 'racking'.

that is, to give it more clarity and lightness before it is put into the barrels, bottles or cans. Some ales are subjected to a secondary fermentation process in the barrels in which they are sold, but this is now comparatively rare, since most beer in bulk is now supplied under pressure in aluminium or stainless steel kegs.

The brewing process yields several useful waste products. Animal feedstuffs are made from the dried rootlets and spent grains of the malted barley, and from the yeast residues, which are also used in human foods, pharmaceuticals, and vitamin concentrates (yeast is a rich source of B-group vitamins). Spent hops can be used as fertilizers, but hops as such are gradually being superseded in brewing by hop extract in powder or pellet form, which leaves no major residues. The main constituents of the beer itself are: carbohydrates (5%); protein (0.6%); small amounts of riboflavin, niacin, and thiamine which are forms of vitamin B; traces of calcium and phosphorus; from 2% to 6.5% alcohol; and up to 90% water. One pint (0.56 litre) of beer contains about 280 calories.

MODERN TRENDS Apart from the sheer size of production, the trend in the second half of the 20th century is towards much larger, often international, brewery groups, and more widespread exporting of beer. The use of highly developed equipment, notably electronic devices, has taken brewing a long way from being a simple batch operation towards a truly continuous process, but full automation has not yet been found practicable for a variety of reasons, notably the difficulty caused by the blocking of filters.

BOOKBINDING

Bookbinding holds pages together and protects them. In the printing industry the types of binding used can be broadly categorized as letterpress binding for books which are held open while being read; and stationery binding for blank or ruled books which lie flat while entries are made. Within the category of letterpress binding there are three different styles: library, edition and perfect binding.

Even in the 1970s, no bookbinding processes except perfect binding are fully automated. Rather they are composed of a number of ponderous automatic and semi-automatic machines which have emerged as mechanized variations of the centuries-old handcraft work. In many bookbinding operations, a large amount of manual labour is still employed.

LIBRARY BINDING This type of binding uses cloth or leather covers and is designed to stand up to constant handling. Books, magazines and so on are not printed one page at a time but rather with 32, 16, 12, 8 or 4 pages set out on each side of a sheet of paper. Machines quickly fold and slit sheets and deliver them in 64, 32, 24, 16 or 8 page sections (signatures) which are then placed in book sequence into separate hoppers on a gathering machine. These machines, which require a number of hand operators, use rotating grippers or gripper arms to build up the sections into book form on a moving belt at a rate of up to 5000 books an hour.

The gathered sections are flattened in a standing press, and normally sewn together on a semi-automatic machine which, in the case of library binding, attaches three or four tapes for added strength. Saw cuts are made across the spine to roughen it and ensure that the tapes stick to it. Endpapers are either glued or sewn front and back for attaching the board covers and giving added protection.

A flexible glue or polyvinyl acetate emulsion is then applied to the spine using either a hand-fed machine with gluing rollers and brushes or a fully-automated unit with glue applicators and driers. The book is then trimmed on all sides except the back with a three-knife guillotine before passing on to the rounding operation, where it is fed between rollers which give the back a convex curve and the fore-edge a corresponding concave shape.

To further eliminate strain on the binding, the rounded spine is backed. This involves beating the back edges of the sections to form lips (slight projections), curving the edges of the pages so that the book can open flat. A heavy roller or plates on a backing machine today replace the hammer originally used for the manual operation. A modern hydraulic machine with an ouptut of about 350 books an hour combines both rounding and backing operations.

Covers used in library binding are two split boards—a thick and thin layer stuck together except for a strip (or split) to hold the trimmed-down outside end papers and sewing tapes. Once inserted into the split they are glued. Drawn-in work uses solid boards with holes that match the cords holding the sewn sections. The cords are laced through these holes, then hammered flush and fixed in a press. After the cover boards have been fixed the book can be finally covered.

Materials used to cover books include morocco and calf leathers, buckrams, imitation leather, linen and cloth. A fully bound work is all leather or another quality material; half bound is leather backed with leather corner pieces; quarter bound is cloth with a leather spine.

Until 1825 all letterpress binding was carried out in this way—as indeed craftsmen had done it by hand for centuries. That year saw the introduction of the idea of making the covers (or cases) separately, which allowed a far greater degree of automation at both ends of the process. A modern case-making plant includes board and cloth cutters and casemaking machines—sometimes web (reel) fed.

EDITION BINDING Except for book restoration and some limited editions, library binding has been replaced by edition casebinding, or publishers' work. Here the machines are linked by conveyer feeders to give continuous mechanical production, which is possible because the cover or case is made separately. Full automation, however, is not yet possible. The

Above: a 19th century bookbindery. On the left are the presses for flattening the finished books. Other operators are 'ploughing' (trimming) the edges of finished books, brushing glue on a binding, and on the right the man with the hammer is blocking (stamping) a book with gold leaf.

Right: a German bookbindery, 1568.

'I bind books, one and all,
sacred and secular, large and small,
in parchment or plain boards alone,
and clasps with a good lock fitted on,
and buckles, and stamping for decoration;
I trim them too in preparation,
and sometimes I will gild the edges
because it earns me handsome wages.'

Jch bind allerley Bücher ein/
Geistlich vnd Weltlich/groß vnd klein/
Jn Perment oder Bretter nur
Vnd beschlags mit guter Clausur
Vnd Spangen/vnd stempff sie zur zier/
Jch sie auch im anfang planier/
Etlich vergüld ich auff dem schnitt/
Da verdien ich viel geldes mit.

G Der

This sequence shows a mechanized bookbinding method.

Above left: printed sheets being moved through a folding machine by air jets.

Above: bundles of signatures (sections of the book) being gathered and collated. They have to be gathered in the right order by the machine.

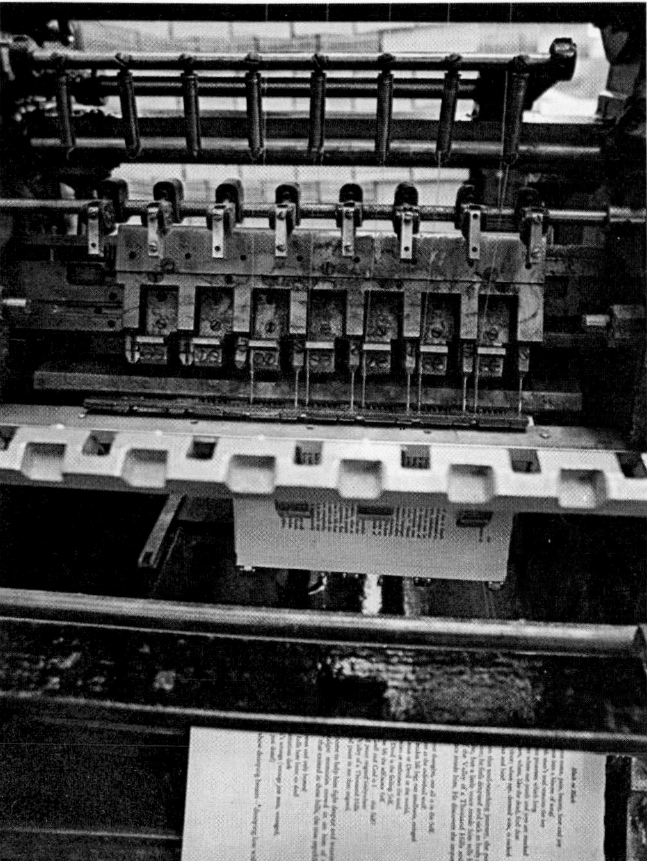

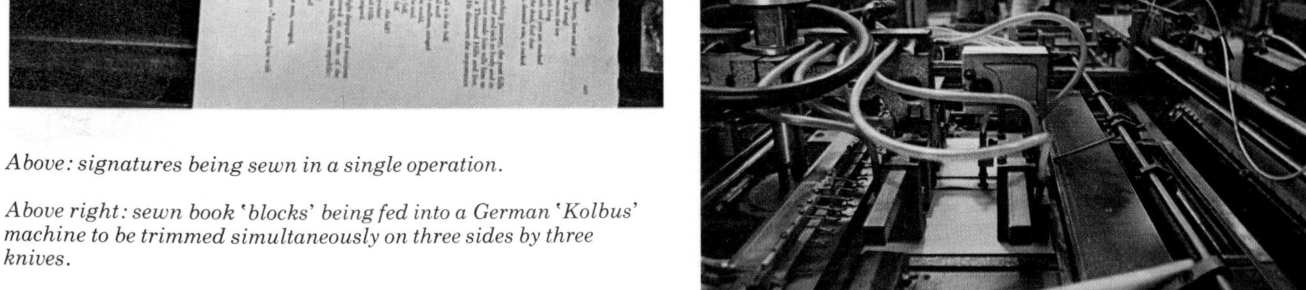

Above: signatures being sewn in a single operation.

Above right: sewn book 'blocks' being fed into a German 'Kolbus' machine to be trimmed simultaneously on three sides by three knives.

Right: a case, or cover, being made, showing the two boards with cover material being attached.

This sequence shows a more traditional method of library binding, with simpler machines.

Right: folding is done on a smaller version of the machine shown on the previous page.

Far right: sewing signatures together. The spine of the book is strengthened by tape fed from the reel at the top. The strong thread comes from several reels in back.

Above: tying off the tapes with thread. The book has been trimmed and saw cuts made on the spine.

Far right: rounding the book. This gives the spine a convex shape; then it is backed by hand with a hammer.

Right: cutting boards to size in a mechanical guillotine. The extending frame keeps the operator's fingers safe.

Right: making cases, or covers. They are entirely handmade. The boards are fastened together by flexible liners, which are slightly wider than the thickness of the book to allow for the convex shape of the spine and the lips, or projections, created by the backing process.

collation and sewing stages of the book are the same as those mentioned under library binding. The book then passes to a casing-in machine which pastes the end papers onto the book; the book rises to the case; its spine meets the spine of the case; the boards fall and contact the end papers. The covered book is then ejected and forwarded through a pressing machine.

PERFECT BINDING Nowadays, some of the most advanced bookbinding machinery is involved in perfect binding, which is the method used for paperback books. This may not be as durable as sewn work but the degree and speed of the automation makes it

Left: casing in, or fitting the covers. In high quality books, the ends of the tapes running across the spine and the edges of the endpapers are inserted in splits in the edges of the boards.

Below: adding a transparent dust jacket.

the most economical. Printed sections move along a conveyor belt and are automatically gathered and collated, after which the back edge is trimmed and roughened. A fabric lining is glued to the back and the paper cover is then wrapped round and glued. The final trimming is performed in one operation by a three knife guillotine. Development work is going on with the aim of making perfect binding good enough for higher-quality work.

OTHER METHODS One semi-permanent method of holding a book together is known as mechanical binding, and is used for catologues and notebooks. The pages are slotted or drilled together along the back edge and held firmly while a metal or plastic wire is threaded through as a comb or spiral. Some plastic combs can be unclipped so that pages can be inserted or removed if necessary.

Other simpler and cheaper methods include wire stitching—saddle stitched and flat or side stitched. Saddle stitching is used on insetted work (sections placed inside one another, such as a magazine). The job is opened at its centre and placed on the saddle of a wire stitching machine. Wire is fed from a reel, driven by the stitcher head through the back fold and clenched underneath to form a staple.

Side stitching is used when separate sections are gathered together. The wire is stitched from front to back through the side on the binding edge, and clenched as in saddle stitching. A variation of this is known as side stabbing, where staples not long enough to pass right through the book are inserted from both sides.

STATIONERY BINDING In stationery binding for account books and ledgers, as all pages or pairs of pages are the same, there is no need for complex gathering and collating techniques. The ruled sheets are simply folded and machine sewn with strong tapes. Because of the greater size and substance of the pages, cotton linings often reinforce the outer sections and the endpapers are thicker than in letterpress binding. The backs are glued and the book is trimmed on three sides before moving onto the rounding and pressing machines. Leather, cloth or canvas back linings are applied, and millboard stiffeners are attached to a spring back, which ensures the pages will stay open and flat. The spring back is a roll of laminated board and paper that virtually grips the back of the book and is attached to the split board covers. Finally the covering material is pasted, drawn on and turned in, and the end leaves are glued down.

CANDLE making

Candle making dates back to the ancient world. Cone-shaped candles were depicted in relief in Egyptian tombs, while dish-shaped candlesticks, dating back to about 3000 BC, were found in Crete. Historically,

An old time candle maker pouring wax into holes in a table through which wicks are suspended.

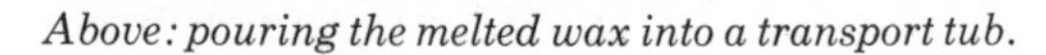

Above: pouring the melted wax into a transport tub.

Above right: putting pieces of solid wax into a chamber where they will be melted by steam heat.

Right centre: candles for churches being made in the old-fashioned way. The beeswax is poured down the suspended wick. The candles burn with a very clean flame.

Lower right: the beeswax candles in various stages. They must be turned end-for-end halfway through to ensure uniform thickness.

These two pictures show the recently developed extrusion method of making candles. The candles are made of crushed paraffin wax which is forced through a die under pressure; at the same time it is consolidated around the wick. The machinery, which was designed and built by Price's Patent Candle Co Ltd of London, then conveys the candles to the packing stage.

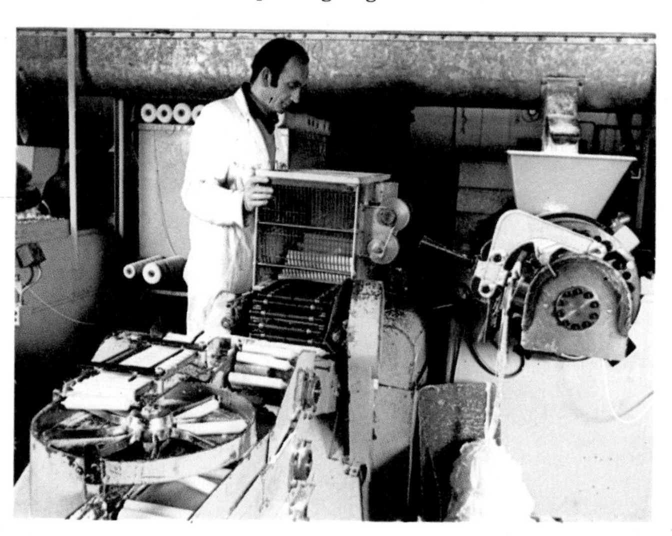

they have been used for a variety of purposes apart from illumination. In the time of King Alfred the Great they were used for timekeeping, and until the present century the bidding time at auctions was often limited by inserting a pin into a candle which eventually fell out as the wax around it began to melt from the heat of the flame. Candles were also used to determine the duration of miners' shifts.

Candles are made from a wick of twisted or plaited cotton surrounded by animal or vegetable fats or paraffin wax refined from crude petroleum. Methods of candle manufacture can be briefly described as dipping, pouring, drawing, moulding and, in recent times, extrusion and compression.

The dipping method, dating back to the Middle Ages, used lengths of wick made from dried rushes which were peeled, except for a strip at one side, to reveal the pith. These were repeatedly dipped into melted fat until enough had adhered to the wick. The candle produced gave a smoky flame and lasted for nearly an hour.

Compared with fat, beeswax was much cleaner and used in wealthier establishments. It is still used in Christian churches. Beeswax candles, sometimes reaching a height of six feet, are still made either by the dipping or pouring method. In pouring, melted wax is ladled over a suspended, plaited cotton wick which is twirled by the fingers at the same time. As it cools, more wax adheres to the lower portion of the wick so halfway through the process, the wick is reversed and pouring continued.

MANUFACTURING TODAY The most widely used method for candle manufacture today is moulding. Hand-operated machines, each containing about 500 tin moulds, are filled with molten paraffin wax and then cooled by water circulated in an outer jacket. On cooling, the wax solidifies and shrinks slightly. This facilitates ejection from the mould, which is done with movable tip pieces attached to hollow pistons through which the wicks are passed from spools beneath. After ejection, when the piston tips have descended through the then empty moulds, lengths of wick are left in position for the next candle. This process is repeated, normally every twenty minutes, and can be used to make cylindrical, tapered or fluted candles provided their shape permits ejection from the mould.

Another method is the extrusion of candles from solid paraffin wax broken into small particles, which was developed by a London candle company in 1950. It marked a big step forward by automating the production of parallel-sided candles and also has economic advantages.

Crushed paraffin wax is forced, under high pressure, through a small orifice where it is consolidated around a wick. This produces a continuous length of candle which is automatically cut into specific lengths. The tips are formed by rotation cutters and the finished candle is conveyed to an automated packing machine.

Cotton wick used for candles is usually plaited and chemically treated so that it bends, in a similar way to a dried rush, to an angle of 90° when burning. In

Left above: the molten wax poured into a moulding table, where it runs into moulds around the wicks.

Above: the candle maker peels a layer of hardened wax off the top of the table. It will be re-melted.

Left: he turns a crank, operating a mechanism which pushes the candles out of the moulding holes.

Below: After snipping off the wicks, which are pulled up continuously from beneath the table, he carries off a tray full of brand new red candles. The table is ready for the next batch.

this way, the end is brought into the hot outer mantle of the flame causing it to be shortened naturally and eliminating the need for snuffing—an irksome task with untreated twisted cotton wicks.

Coloured candles are popular, and some are made to give off pleasant aromas. Some dyes and perfumes, however, have undesirable effects. They can cause the wick to become clogged with unburnt carbon which produces a poor smoky flame, resulting in liquid wax forming in the cup just beneath the flame. This tends to 'gutter' and run down the side, sometimes causing the complete collapse of the candle.

One of the many uses of candles includes night lights, which are made of powdered wax compressed into small blocks with thin wicks. These are inserted into cups made of glass or leak-proof, fire-resistant paper.

CANNING of foods

Canning is the heating or processing of food in hermetically sealed containers so that it will keep for years at room temperature. Nicolas Appert, who experimented in the early 19th century, is generally considered the inventor of the process, although the reasons why canning preserved foods so successfully were not known until Pasteur showed the connection between micro-organisms and spoilage some 40 years later.

Micro-organism—bacteria, yeasts and moulds—are very small and simple life forms present in all natural foods, air, water and soil. Their actively reproducing (vegetative) forms are easily killed at relatively low temperatures—around 176°F (80°C)—but some bacteria form 'resting bodies' called spores, which can survive much higher temperatures.

A wide range of fruits, vegetables, meats, milk, fish and speciality products (like soup and baked beans) are canned. Dangerous spores will not grow in acidic foods and so these, for instance fruits, may be processed at 'low' temperatures (212°F, 100°C). All other less acidic commodities must be processed at higher temperatures (240°F, 116°C or higher).

CAN MANUFACTURE The can is a hermetically sealed container, strongly made to ensure that the contents are kept sterile. For many years the traditional can has consisted of a soldered body, usually cylindrical, with metal ends, one fixed by the can maker and the other by the canner. It is therefore known as an open top or 'sanitary' can, and is used for food products and beverages.

The can maker starts with plain or lacquered tinplate sheets which are slit into smaller sheets or body blanks appropriate to the can size being made. As the blanks pass through a body maker they are flexed and notched at one side; slits are put in the other, and the edges appropriately bent. The blank is formed into a cylinder on a mandrel, and the hooked edges are engaged and hammered to give a mainly locked seam with short lap seams at the extremities to facilitate end fixing. The cylinders are then fluxed and preheated to make them easy to solder, and carried by magnets or grippers known as 'dogs' over solder rolls, heated to ensure penetration of the solder into the seam. They are wiped to remove surplus solder and finally cooled. The top and bottom edges of the cylinders are flared out to provide flanges. The can ends are stamped out on a press and the perimeter bent under to help the later seaming stage. Finally the ends may be lined with a synthetic rubber lining compound. The can is made by interlocking the flanged cylinder with the end by a spinning process, known as double seaming, with such precision as to give a very tight closure.

The aseptic canning process is suitable for thin foods, such as soups. The food is cooked, sterilized, cooled and transported in the tube system, then piped directly into pre-sterilized cans and sealed.

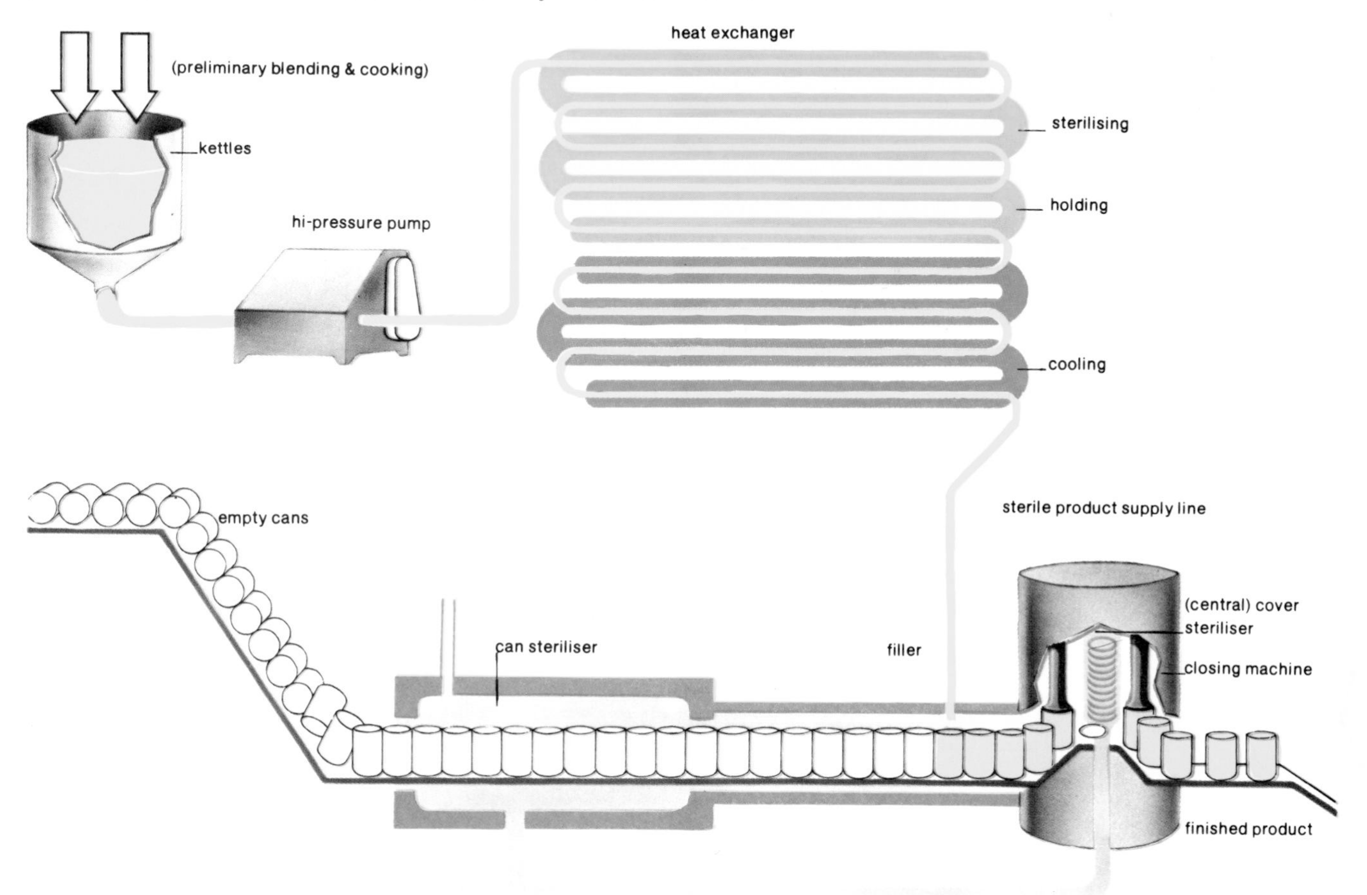

Above: an automatic strawberry canning machine. The cans are filled to the top with fruit and some syrup, but often when the can reaches the consumer it seems to contain mostly syrup. This is because the fruit contains a lot of moisture, which is gradually drawn out into the syrup by a natural process called osmosis.

Shallow drawn containers in tinplate and aluminium have also been used for a long time, particularly for fish products. These are stamped out on a press and the flange trimmed. They may be cylindrical, oval or rectangular, with straight or tapered sides, and the end is fixed by the canner by double seaming it on to the flanged body.

Most cans are made from tinplate. This material is basically mild steel strip which has had a protective layer of tin added, usually by electro-deposition, to prevent the steel from rusting. Aluminium and other forms of steelplate are also used.

DEVELOPMENTS Since some 60% to 70% of the total cost of a can is in the material, it has been important to maintain its competitive position by the introduction of technically proven economies. Thus the thinnest possible tin coating is achieved by differentially coating the plate on each side to the level required. Chemically treated steel plate, usually chrome or chrome oxide protected, known as TFS or tin free steel, has been introduced. Container thickness has been reduced in some instances by using stronger double reduced plate, which is produced by the steel manufacturers by an extra heavy final rolling technique. Modification of can design, for example, by the introduction of beaded bodies, which are ridged for extra strength, has also contributed to cost savings.

In the manufacturing process, the utilization of tinplate and efficiency and speed of the can making operation have been highly developed. Perhaps one of the more spectacular examples is where a long cylinder made on a standard bodymaker is divided on a special parting machine into two or three separate cylinders, doubling or trebling the speed of the line.

Below: cans of soft drinks on a conveyer which divides them into two rows for packing into cartons.

Below: this machine seals 500 cans per minute. A space left above the drink is filled with carbon dioxide gas before the can is sealed, which eliminates airborne bacteria.

2

3

Welded or adhesive bonded seams are being used increasingly for beverage cans. Such seams are even stronger than the metal itself. The adhesive used is a polyamide type and is pre-applied to the specially lacquered tin free steel plate before the blanks are heated and formed into cylinders. On the lap side the seam is bumped together with highly chilled tools to set the hot adhesive. It is most important that the cut

The latest development in can manufacture is the drawn and wall ironed technique used widely for making beer and soft drinks cans. On the wall ironing line, tinplate is fed from the roll (1) into the cupping press, seen in the background, where it is tested, using ultra-violet light, for pinholes in the plating as well as evenness of the tinplate. Six discs of metal are blanked out in the press at a single stroke (15,000 in an hour), and are drawn into half size cups (2). The waste tinplate is recovered for further processing. The half size drawn cans are then passed to the large wall ironing machine (3) where they are stored overhead in large bins before being fed via a conveyer system into the machine to be drawn up to full size.
The cup is drawn up to full size by forcing it with a punch through a number of rings (4). The clearance between the punch and the rings becomes progressively less, ironing the walls thinly while leaving a thick base. The cans are then passed to the washer (5) where they are given two detergent washes and dried. (Sequence of pictures continued on next page.) (6) The labels are printed by litho offset. (7) Necking and flanging the top of the can in preparation for lid attachment later.

5

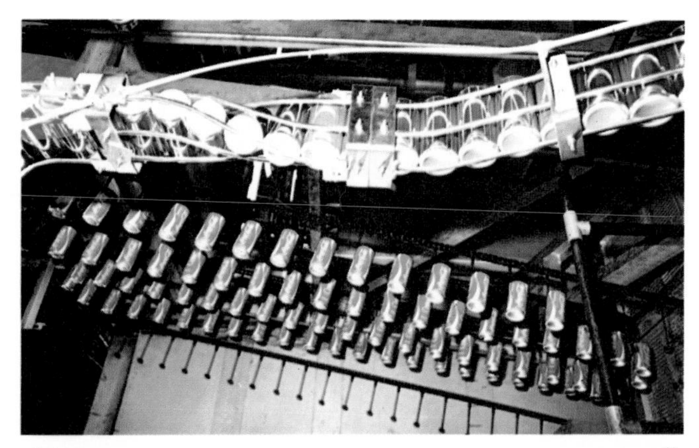

6

7

edge of the side seam is protected either by the bonding material or by applying a side stripe.

The latest commercial development is that of the two piece drawn and wall ironed can (DWI) for beer and soft drinks, made in aluminium or tinplate. This process basically consists of blanking out a disc from a sheet of metal, drawing a cup-shape in it using a press, finally forcing it with a punch through a number of rings. The clearances between the punch and rings become progressively less, thus ironing the walls to a thin section but leaving a thick base. The can is then given a reduced neck form at the top and flanged to allow an aluminium easy-open end to be seamed on.

Cans must be lacquered internally to prevent contamination of the contents. On seamed cans the lacquer coating is applied by roller to the tinplate before the can is made, and on the DWI type of can the lacquer is sprayed on after drawing. Welded seams, found on some TFS beverage cans, need careful lacquering to cover the internal cut edge and welding, which is done electrically using copper wire electrodes to avoid tin contamination of the electrodes.

Below: A, the traditional seamed can making process. The tinplate body blank has already been printed, notched and slit along one edge in preparation for seaming. B, after being formed into a cylinder the body is ready for seam soldering. C, at this stage one end is attached.

A

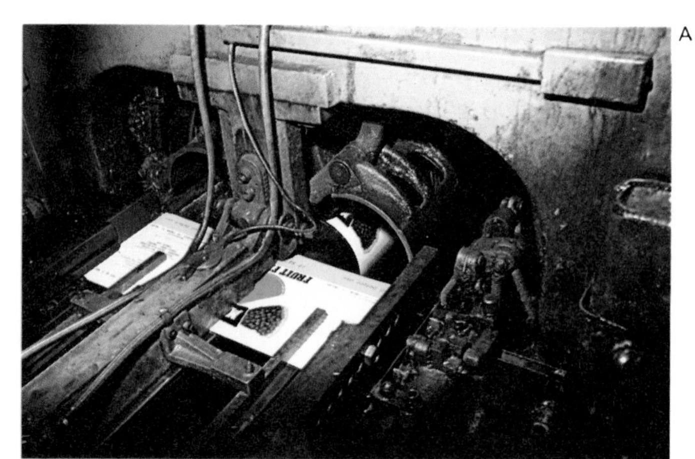

B

C

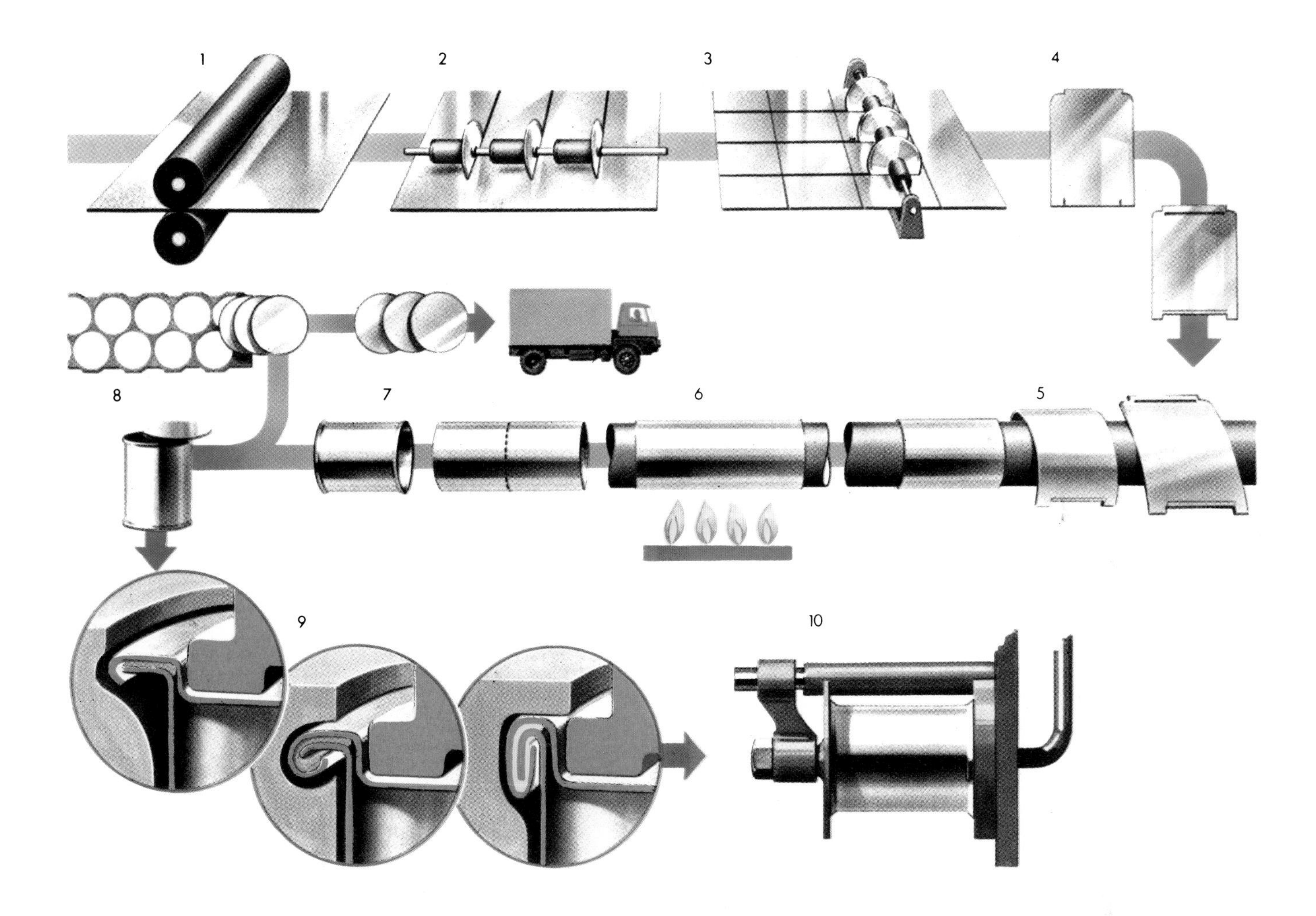

The tinplate is laquered and may be printed. (1). The sheets are split (2) before being cut into body blanks (3). In the bodymaker (4) they are notched along one side, slit along the other and a seam hook is formed. The body blanks are formed into cylinders (5) and the side seam is soldered (6). Each end of the cylinder is flanged outwards (7). Can end blanks are punched out (8). One end is attached by a spinning process to form a double seam (9). Finally cans are pressure tested.

Many billions of cans are manufactured annually throughout the world. In Britain alone, about 8500 million are produced every year.

CANNING OPERATIONS Most operations are mechanized on a production line basis. The machinery used, especially in preparation, depends upon the nature of the raw material. The aim is to pack the cleaned and edible portion of the food. These operations, such as peeling root vegetables or trimming meat, are essentially those used in domestic preparation.

Blanching, a short preliminary heating at 180 to 212°F (82 to 100°C), is important for killing the enzymes which can cause unwanted effects such as discoloration or odd flavours; it also removes air from the product, and aids filling. Prepared and inspected material is put either manually or automatically into pre-washed cans. In multi-stage operations, solids are filled first and then topped up with 'liquor' or sauce leaving about a $\frac{1}{4}$ inch (6 mm) headspace which prevents excessive pressure forming during processing.

The vacuum in cans is obtained mainly by hot filling, but may be assisted by passing cans through a shallow hot water or steam tunnel. This produces water vapour in the headspace, so that when the can is cooled after sealing the vapour condenses leaving a partial vacuum. Alternatively, the headspace is swept with steam just before closing, or the can is seamed in a vacuum chamber. The process of creating the vacuum is called deaeration.

PROCESSING Cans are processed to destroy all dangerous micro-organisms capable of growth in the particular food by either batch or continuous methods. In the former, crates of cans are loaded into retorts which are similar to large domestic pressure cookers. The lid or door is clamped tightly and steam introduced to 'purge' all air from the retort through a vent pipe. This is then closed and the vessel is pressurized and heated to the processing temperature which is maintained by a temperature controller. After heating for a predetermined time the cans are cooled by admitting chlorinated water while initially maintaining processing pressure with compressed air to avoid straining the cans. Retorts are operated at 212°F (100°C) for fruit (acid) packs and up to 260°F

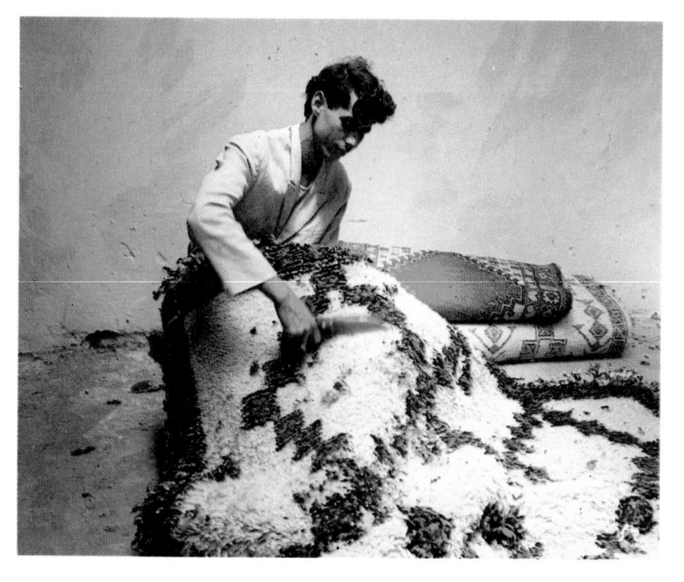

Top: Tibetan carpet makers at work on a vertical loom. The carpet at the top of the picture is complete, and is hung up as a pattern for the weavers to copy. They start at the bottom; when the working edge gets above a comfotrable height the carpet is wound around the lower beam.

Above: hand weaving gives carpets an irregular pile, which must be cut to an even length. This is a Berber carpet from Morocco.

(127°C) for low acid products. Continuous cookers, capable of processing 2000 cans per minute, may be operated at up to 270°F (132°C). In these machines, cans are introduced into the steam chamber through mechanical valves. Hydrostatic cookers use high columns of water instead of valves to balance the steam pressure.

RECENT DEVELOPMENTS Free flowing products, such as vegetables in brine, may be heated directly by rotating the cans over gas burners in flame cookers without damaging the contents. The system requires heavy cans with strengthened ends to withstand the high internal pressures which develop within them during heating.

In aseptic canning, thin films of homogeneous products are heated and cooled. Compared with conventional canning, much shorter process times and higher temperatures—up to 300°F (149°C)—can be used to sterilize the food. The cooled, sterile product is filled into pre-sterilized cans which are seamed in a sterile atmosphere. This method allows sensitive products like custard to be processed without developing overcooked flavours.

Recent advances in plastics technology have resulted in laminated plastic containers which incorporate an aluminium foil layer to provide a gas and moisture barrier. These containers are capable of withstanding the high temperatures required to sterilize foods. The most widely used type is the flexible pouch. After filling, the pouch is closed by fusing the inner plastic layer of the laminate in a heat sealer and then it is processed in a retort. Flexible pouches may be used for a large number of products and are expected to extend the range of heat processed foods available to the public. They are already being processed in increasing numbers in Japan and Europe and have been used by American astronauts on the Apollo programme.

CARPET weaving

Carpets are a comparatively recent innovation in the Western world: few were seen before the eighteenth century, and they did not become widespread until the invention of mechanized carpet looms in the nineteenth century.

Carpets had, however, been made on hand looms in the East for a very long time. The oldest carpet known is the Pazyryk carpet, which dates from about 500 BC. It was found in a tomb in the Altai mountains of Central Asia: the log-lined chambers had filled with ice, preserving their contents perfectly.

HAND MADE CARPETS The Pazyryk carpet is Persian, and was made using the *Ghiordes* or Turkish knot, still the most common for hand made carpets. These are made on looms, as for simple woven materials, but instead of one yarn (the weft) simply being woven in and out of the other (the warp), a third yarn, to make the pile, is included. This is done by knotting the pile yarn on each warp thread in turn, then threading through the weft in the normal way, usually twice or three times for each row of knots. The weaving is then pushed down with the fingers or a comb, so that the knots are firmly trapped by the weft. Another row of knots is then made, and so on.

The knots used, such as the Ghiordes and the *Sehna*, the most common types, are designed so that both ends point outwards, taking in one or two warp threads for each knot. By choosing different colours of pile yarn, the intricate patterns of carpets can be built up. The pile is cut to a uniform length when completed.

The Ghiordes knot is the coarser of the two, with a longer pile, and can only be used for straight line patterns. Any attempt at a curve gives a stepped, ill-defined appearance. The Sehna knot is more secure and can therefore be used for a shorter pile with a finer mesh, allowing a curved design.

CARPET TYPES The size of the mesh can vary widely, a typical value being about 250 knots per square inch (40 knots per cm^2), in which case the knots are about 1/16 inch (2 mm) apart. Some carpets may have millions of knots, taking several years to make.

Oriental carpets and rugs can be classified according to their pattern and the way they were made. Nomadic tribes make carpets on portable horizontal looms which can be dismantled when moving from camp to camp in search of new pastures for the sheep. These carpets usually have bold geometric patterns with vertical, horizontal and diagonal lines. Flowers, animals, insects, stars, and so on are rendered in geometric terms, the designs being handed down by word of mouth. The works nomads produce are usually fairly small rugs, the size being dictated by tent and saddle requirements, and are made of wool with a little camel hair and silk added.

Rugs with more conventional patterns are made by village weavers who live in comparatively isolated communities. Men, women and children weave during the winter months on vertical looms. These may have the warp stretched between upper and lower fixed beams, in which case the rugs are limited in length, or they may have a lower beam which can be rolled up or moved so that the length of the carpet can be extended.

These village carpets are often stylized, with repeated geometric patterns, more subdued than the nomadic designs. Because the looms are more permanent the carpets are longer.

The most striking oriental carpets are those made on a large scale in court or studio workshops. These often have intricate floral designs, and may be up to 40 ft (12 m) in length. Many of these carpets, which date back to the sixteenth century, are impressive works of art. Fine wool, silk, gold and silver threads were woven by nimble fingered children who were particularly adept at weaving the fine intricate designs. To guide the weavers, a *wagireh* of squared paper, where each square represented a knot, was used.

Below: carpets are woven on looms, like any other fabric, and the warp (vertical) and weft (horizontal) threads are shown here in a pale colour. As well as the normal weft, pile threads (pink) are also knotted in, their tops (brown) forming the surface. Two knots are used in most handwoven carpets: (below) the Ghiordes (Turkish) and (bottom) the finer Sehna (Persian) knot.

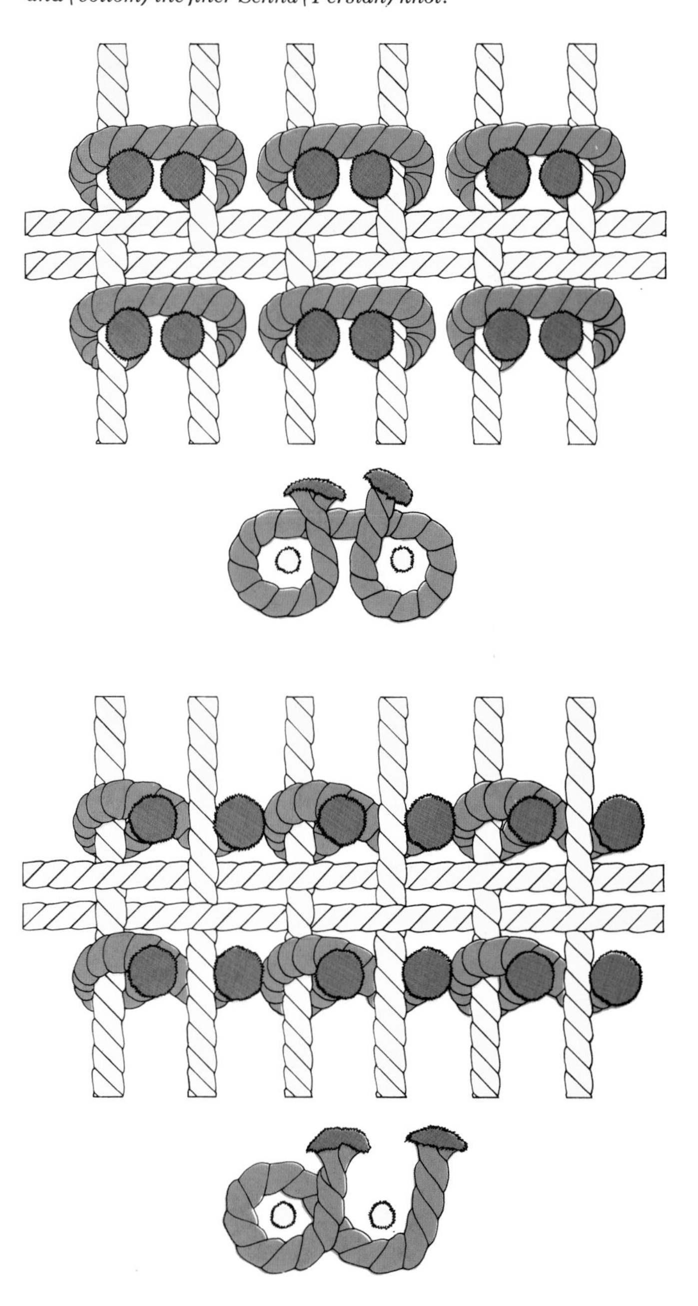

Not all carpets have a pile. The Turkish *kelim*, for example, is a straightforward woven fabric, with no knots. In this case, designs are produced by the weft colours; instead of the weft always going from one side to the other, it is doubled back when part way across, so producing small patches of a colour.

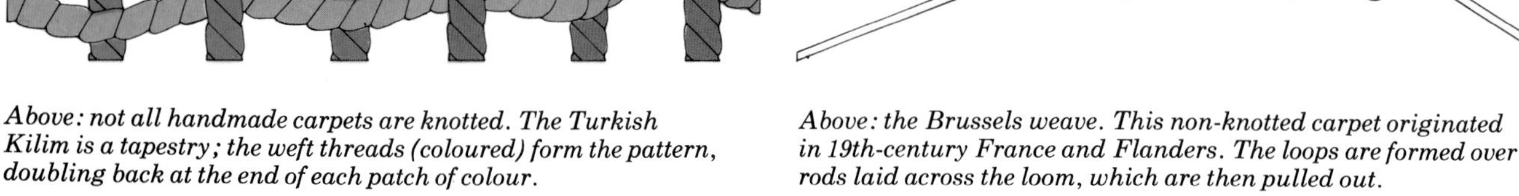

Above: not all handmade carpets are knotted. The Turkish Kilim is a tapestry; the weft threads (coloured) form the pattern, doubling back at the end of each patch of colour.

Above: the Brussels weave. This non-knotted carpet originated in 19th-century France and Flanders. The loops are formed over rods laid across the loom, which are then pulled out.

Below: the Wilton carpet is a development of the Brussels weave (above right). Loops of pile are woven in around blades which then rise to cut them into tufts.

Below: in the Axminster carpet, tufts are inserted individually, either from spools or by means of a gripper mechanism. This is the structure of a gripper Axminster.

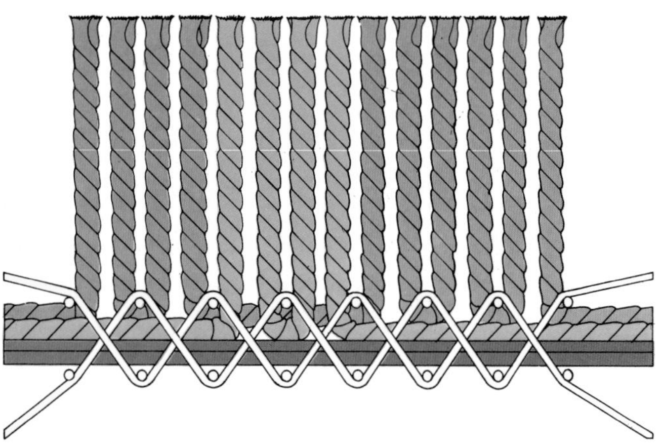

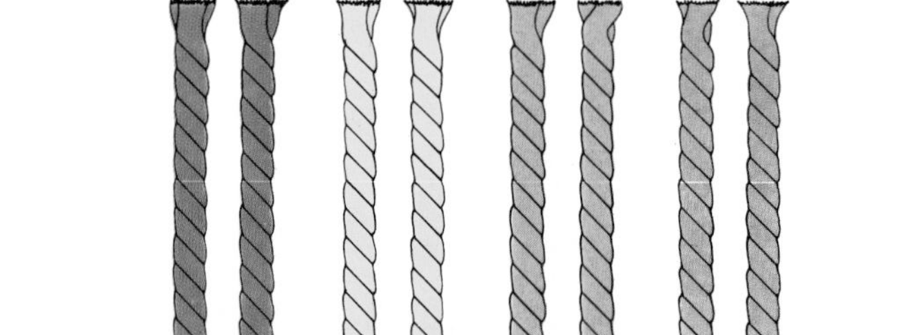

DYES The often brilliant dyes are produced from local vegetation, such as madder root (red), soumak (yellow and brown) and indigo (blue). Pomegranates, walnuts and other fruits also give stable dyes when fixed with a mordant. These are mineral or vegetable substances which react chemically with the dye forming an insoluble compound which attaches itself firmly to the yarn. Typical mordants are alum, tin oxide and iron sulphate.

EUROPEAN DEVELOPMENTS The art of carpet making by hand spread from the east into Europe, where it became established in France and Flanders. A new type of non-knotted weaving developed, known as Brussels; the principles used in this turned out to be suitable for mechanization.

The pile in a Brussels carpet is formed of loops rather than free ends. The pile yarn is looped over rods laid across the loom during manufacture and later withdrawn. In the simple knotted carpet, there is only one layer of warp and weft, but in the Brussels carpet there are two layers, with two shots of weft sandwiching an extra pair of yarns called stuffers, which strengthen the fabric.

WILTON CARPETS Towards the close of the seventeenth century, persecution of the Huguenot Protestants on the continent forced them to flee to England. Some settled in Wilton, Wiltshire, where the Wilton carpet was developed. This uses a Brussels weave, but with the rods replaced by blades which cut the loops to form tufts as the blade is withdrawn.

The invention of the Jacquard loom in France in 1801 made it possible to weave Brussels and Wilton carpets in up to five colours. Each colour of pile is used over the whole area of one carpet, but only the colour which is to be seen at any one spot is brought out as a loop. This extra layer of pile, though wasteful, gives a luxurious thickness to the carpet.

Wilton carpets are still made on large mechanical Jacquard looms. One type of machine weaves two carpets at once, face to face, with the loops going

from one half to the other. A single blade is used to separate the two and cut the tufts.

AXMINSTER CARPETS In 1755 a cloth weaver called Thomas Whitty in Axminster, Devon, set up a factory producing hand knotted carpets of fine quality. This led to the use of the word Axminster to describe many different carpet processes in which the tufts were inserted individually rather than as a continuous weave.

Both processes in use today for making Axminster carpet were based on inventions of Halcyon Skinner of Yonkers, New York, but were developed by firms in Kidderminster, England. The spool Axminster machines have each colour required for the pile of the

The four pictures on this page and the first two on the next show the manufacture of gripper Axminster, in which a break-like device called a gripper takes a tuft of pile yarn from a spool and inserts it at the appropriate place. The selection of yarn is controlled by a Jacquard mechanism, one of the first automatic programming devices to be invented, where the pattern is coded on punched cards.

Above left: the row of grippers swings up and each pulls out a length of pile yarn.

Above right: they fall back, carrying the cut yarn, and insert it in the carpet below.

Below left: the emerging carpet is carefully inspected: a malfunctioning gripper would cause a 'run' of missing threads.

Below right: behind the loom, a stack of spools feeds the pile yarn to the carriers from which the grippers pull it out. In the foreground, a man refills a spool.

carpet (in theory an unlimited number) pre-loaded side by side on horizontal spools the width of the carpet. One spool is prepared for each line of pile, and the spools are arranged one above the other on an endless chain so that after a while the pattern is repeated. A typical number of spools is 288, with the pattern re-occurring every yard or metre.

Each spool has a row of small tubes along its length, with the ends of the pile yarn threaded through them and projecting a short distance from the end. A pair of arms take one spool at a time from the chain down to the weaving point, where the projecting ends stick through the warp threads. The weft is then shot through twice, trapping the ends; these are then turned back, and another weft shot through, securing the turned back end. The spool is lifted slightly, to pull more pile yarn out, and a blade cuts off the projecting end which is now woven into the carpet. This process is repeated for each line of carpet.

In the gripper Axminster mechanism, a smaller number of different colours, usually eight, is possible. Instead of a separate spool for each line there is a metal strip called a carrier, with a row of eight holes through which each colour of pile yarn, loaded on bobbins, projects. There is a row of carriers, each one offering a choice of eight colours, across the width of the carpet, with one carrier for each line of tufts in the finished carpet—usually about seven per inch (2.5 cm). For each carrier there is a gripper. These are beak-like mechanisms, arranged side by side in a long row, and moving together on an axis along the row. The grippers 'peck' a length of each yarn from the carrier, the yarn being chopped to a length of about $\frac{3}{4}$ inch (19 mm) as it is pulled away from the carrier. The correct colour is selected by the Jacquard loom punched card mechanism by means of moving the carrier up and down.

The grippers transfer the pile to the weaving point, where they are woven into the carpet as in the spool Axminster method. Another version, the spool gripper, combines features of both types, with grippers taking the pile directly from spools.

Axminster carpets are more economical on pile than Wiltons, since none of the pile is lying 'dead' beneath

Above: a rear view of the loom, showing the lines of pile yarn leading into the back of the machine.

Right: general view of the loom, showing the punched cards for the Jacquard mechanism at the top. These select the pattern which can be of great complexity provided it does not use more than eight colours. The cards circulate in a continuous loop.

Next page, top: the upper part of the picture shows the design for an Axminster carpet made on the machine previously shown; under it is a section of an actual carpet made to that design.

Far right: this loom makes tufted carpet, an inexpensive type where the pile is inserted into a backing fabric, but not knotted. This is a 'sculptured' carpet, with different lengths of pile making a pattern.

the visible part, though this is often compensated for by making the pile thicker. They can be cheaper, and are therefore more widely used.

TUFTED CARPETS A large share of the market for carpets has been taken by the tufted carpets. These put soft floor-coverings within the reach of many more people, and led to an enormous increase in carpet production. Tufting was evolved in the 1920s from the conventional sewing machine, and involves stitching tufts to a previously woven fabric. A needle inserts the pile from what is to be the underside of the carpet.

It is held by a hook-shaped looper, which has a knife at one end to sever the loops and make the pile. To hold the pile in place, a coating of rubber latex is spread over the underside. A layer of hessian material is often added for stability, which can give the impression that the carpet is woven.

A variety of techniques for producing carpets are based on sticking fibres or tufts onto an adhesive-coated backing. In one process, flocked carpets are made by directing chopped fibres of uniform length onto an adhesive-coated backing, using an electrostatic field to project them and maintain their upright position to form a pile. Another method, producing floorcoverings which are popular for shops and offices, uses a bank of small barbed needles to punch fibres into a woven backing. After this the material is treated with acrylic resin and passed through rollers to consolidate the fibres.

Above: dry ingredients are mixed separately with water to make the liquid 'slips'.

Left: a filter press then forces most of the water out of the slip mixture, forming a 'filter cake', shown here being removed from the press.

CHINA and porcelain

The history of pottery spans some 10,000 years, and during most of this period the clays in use throughout the world produced articles that were predominantly reddish brown in colour. By at least 1000 BC, however, the Chinese had discovered a way of making fine white ceramics using a clay now known as kaolin or china clay. This technique did not spread to Europe until the middle of the 18th century, when the china industry became established in England and Germany.

In common usage today, the term 'china' covers all white vitrified domestic ware. Although the ingredients of china, which together are known as the 'body', vary somewhat from country to country the major one is the white-firing china clay. The exception to this is bone china, which consists of about 50% bone ash (calcium phosphate produced from burnt animal bones).

Whatever the specific formula of the body the way it is prepared is fairly standard. The non-plastic ingredients such as quartz and Cornish stone (felspar or china stone) have to be reduced to particles sufficiently fine to be mixed with the clay. This is done by grinding them in rotating or vibrating mills lined with ceramic brick to avoid metal contamination of the ingredients. Small balls of very hard porcelain may be put into the mill to assist the grinding process, and these are separated from the ingredients after the milling.

Each of the ingredients has an important role to play. The quartz, for example, reduces shrinkage during drying and firing; the Cornish stone acts as a flux. The fineness of the particles is important and often these ingredients are purchased ready-milled. The clay is normally supplied as dry lumps or powder rather than in its plastic form, and other white-firing clays such as ball clay, which is similar to china clay, may be used in addition to the china clay. In some sanitary ware, for example, half of the clay used could be a white-firing variety other than kaolin.

MIXING To ensure complete mixing of the ingredients it is usual to suspend each one in water, the resulting liquids being known as slips. Each slip has its own tank or blunger, and is continuously agitated to prevent sedimentation.

These slips are pumped into a mixing vat, called an ark, in the desired proportions. Typical proportions for bone china are 50% bone ash, 25% clay and 25% stone. Other chinas may contain, for example, 40% to 50% clay, 15% to 25% quartz and 20% to 30% stone, but the proportions vary according to the type of ware being made. Sometimes the non-plastic ingredients are added to the ark in powder form.

The next stage is to turn the liquefied body into a plastic form that can be shaped or moulded. This is done by large filter presses that force out a certain amount of water through filter cloths. The resulting 'filter cakes' are sheets of plastic body, usually about two feet (60 cm) square and two inches (5 cm) thick.

These cakes are not yet ready for use, however, as they contain trapped air bubbles and are wetter on the inside than they are on the surface. To prepare them for use they are fed into pug mills which are really giant mincing [grinding] machines that mix up the cakes in a vacuum to remove the air, and extrude solid cylinder of clay.

MOULDING The oldest known method of shaping clay is by hand alone; a later invention of profound importance was the potter's wheel. Both these techniques are slow and unsuitable for producing articles of uniform size and thickness. In a modern china works, most of the shaping is done automatically or semi-automatically.

To make flatware—a plate for example—a measured quantity of clay is rolled out to a predetermined thickness rather like a piece of pastry. This is placed on a mould that shapes the face of the plate, and gently pressed against it by a tool profiled to follow the contours of the mould. The result is a plate of even thickness resting on the mould.

The next stage is the drying. It is here that the mould plays another essential role, as it is made of plaster of Paris and can thus absorb water. The moulds are slowly conveyed through a drying oven and emerge with the ware sufficiently dry and shrunken to be removed, and strong enough to have any rough edges tidied up, or fettled, prior to firing.

Cups are produced by forcing a lump of clay against the sides of a cup-shaped mould with a rotating tapered roller. Hollow wares that are not symmetrical cannot be made in this manner and are formed by slip casting.

SLIP CASTING To make a complex shape such as a teapot a plaster mould is prepared and cut in half lengthwise. The halves are temporarily joined together and slip is poured into the mould. As the plaster of the mould begins to absorb the water the slip at the sides begins to thicken, and a thin wall of clay is built up. After a suitable time the excess slip is poured out and the mould put into the drier. When dried, the mould is opened up and the pot, in its unfired or 'green' condition, is removed. The handle is cast

Below: the filter cakes next go to a pug mill, which uses a vacuum to remove air bubbles.

Right: a plate making machine takes slices from a clay cylinder and presses them into shape.

Above: separating the halves of a plaster mould for a slip-cast teapot, dried before firing.

Below left: dipping 'biscuit' ware into a liquid glaze. Next, it is put in the kiln on the right for the 'glost' firing at 1000°C (1830°F).

Below centre: high-quality ware is still often hand-painted; this is Minton tableware.

separately and fixed in position with a little slip as the adhesive. When completely dry the pot is ready for firing. Other items of domestic ware such as basins, sinks and toilets are made by the slip casting method.

FIRING Early pottery was simply left to bake in the sun, a practice which produced vessels that were quite rigid enough to store dry solids but which were porous and easily reverted to their plastic state when wet. Baking the wares over a wood fire, or, later, in a wood fired kiln was a great advance, as they would remain permanently hard. The porosity remained, and the vessel would always be damp, but the pot would not become plastic and liquids could be stored in it. An example of such pottery is terracotta, of which some garden wares are still made.

The Chinese, however, discovered many centuries ago that given a high enough temperature over a sufficient period of time clay would vitrify, that is, its physical nature would alter so as to render it non-porous. The control of the firing temperature was quite difficult, but modern kilns with gas or oil burners are relatively simple to heat up and to control and vitrification is quite easy to achieve.

The modern kiln is usually a tunnel up to a hundred yards long, with a carefully controlled temperature gradient: the temperature increases gradually along the tunnel, reaching its maximum in the middle and decreasing steadily towards the exit. This means that the ware reaches its maximum temperature in the middle of the tunnel, and is cooling off as it emerges.

Each item has to be carefully placed on a kiln car which is then slowly drawn through the tunnel. The setting of the pieces on the car calls for skill and experience as they shrink by about 12% as they are fired. A cup, for example, is set on a cone shaped piece of refractory material so that it can ride up freely as it becomes smaller. Bad setting will cause distortion and cracking.

Another type of kiln in common use is the intermittent one. Unlike the tunnel kiln it does not burn continuously and it is switched off when firing is complete.

Vitrification requires a temperature of about 1200°C (2190°F) and the firing cycle takes between 14 and 30 hours. What emerges from this first or 'biscuit' firing is ware that is non-porous, white and mechanically strong, and ready for decorating and glazing.

GLAZING A glaze is a thin layer of glass completely and evenly covering the biscuit ware. It is applied as a liquid by spraying or hand dipping, the suspended solids melting sufficiently in the second (glost) firing so as to form a skin of glass. The glaze adds a shine and smoothness to china and, in the case of a body that is porous, renders it impermeable.

The composition of a glaze has to be carefully matched to suit the body. Both the glaze and body must, when cooling, shrink to the same final extent so that the layer of glaze fits the body properly. A bad fit can cause the crazing that appears on older pieces of pottery. The glaze may be either clear or coloured, and

Above: chinaware shrinks considerably during firing, typically by about 12%. This picture shows the same figurine before and after firing.

Below: a kiln car loaded with sanitaryware emerging from a tunnel kiln. Sanitaryware is made by slip casting and often decorated with coloured glaze. Glazed china is ideal for these items, since it is strong, relatively cheap and its smooth surface is easy to keep clean.

Above: porcelain has been made in the Far East since well before the 10th century AD, though it was not until the 18th century that a process was discovered in Europe for producing ware of the same quality and hardness. This is an example of 18th-century Japanese porcelain, an ewer with a dragon design made during the Togukawa period. It is of the type called ko-Kutani ware, made from a clay discovered in the previous century by Goto Saijiro at Kutanimura (hence the name).

Left: much modern porcelain is cast in plaster moulds in the same way as ordinary china. What distinguishes porcelain is not how it is made but the ingredients used and the firing temperature, which reaches 1400° C (2550° F).

Right: to reach the extremely high temperatures needed for firing porcelain, great care must be paid to the design of the kiln, and it was this that was the chief reason why it took so long for true porcelain manufacture to begin in the West; the other was that the early European 'soft pastes' were so soft that they used to collapse in the kiln. Chinese and Japanese technology reached an advanced state far earlier than that of Europe, which was still in the Dark Ages when the first porcelain was made in China. The illustration shows a Chinese porcelain kiln, small but effective.

a common use of coloured glazes is in the manufacture of sanitary ware.

DECORATION Decoration is ideally applied before glazing, so that the pattern is protected by a hard layer of glaze. Many of the colour materials used in decorating will not, however, stand up to the temperature of the glost firing, which is a little over 1000°C (1830°F). Thus the decorating may have to be done after the glazing, necessitating a third or 'enamel' firing at about 750°C (1380°F) in order to fuse the pattern into the glaze. The two basic methods of decoration are called under-glaze and on-glaze.

The oldest method of applying colour is by hand, which requires considerable skill as the raw materials used often change colour during firing and this must be allowed for when the colours are chosen. This method is expensive.

The usual way of decorating tableware is by means of paper transfers or lithos. These are printed from lithographic plates, using varnish instead of ink. The powdered colours, usually metal oxides, are dusted on and stick to the varnish. Normally a manufacturer will buy lithos made to his designs by a specialist. The lithos are wetted and carefully pressed onto the ware by hand. The silk screen printing technique can be applied to pottery too, either direct or by transfer.

Some of the most expensive tableware has patterns enriched by hand painting, and gold and platinum is often used in the decoration. The powdered metal is mixed with resin and applied by the normal on-glaze methods or just with a brush. The resin burns out, leaving the metal on the glazed surface. In really elaborate pieces several enamel firings may be needed to bring up the various colours.

PORCELAIN Porcelain may be defined in the modern European terms as a vitreous (glassy) ceramic body, white in colour and translucent. This definition can be enlarged to include certain early Chinese wares, and any Chinese ware sufficiently highly fired and giving a high ringing note when sounded (struck lightly).

Porcelain in China developed gradually out of the old tradition of stoneware, the first examples of which came from the Western Chou Dynasty (1028 to 772 BC). Further examples of progress come from the Han Dynasties (206 BC to 220 AD). These hard fired ceramics with high temperature (over 1250°C, 2282°F) glazes, in turn developed into porcellanous ware, illustrated by the fine examples of Yüeh Celadon glazes (907 to 959 AD). The final development of white porcellanous ware came in the T'ang Dynasty (618 to 906 AD), which was followed in the Sung Dynasty (960 to 1279 AD) by true white translucent bodies and high temperature glazes of various colours.

The art of porcelain making spread to Europe by two different routes, the hard paste method arriving via the sea trade between Europe and China, and the soft paste method arriving in Europe by way of the Islamic empires.

'True' porcelain was first made in the T'ang Dynasty by mixing china clay (*kaolin* or *pai tun*) with *petuntse* (the less fully decayed felspar and quartz). The typical body compositon was 50% kaolin and 50% petuntse (25% quartz and 25% felspar). The hard paste porcelain is made by firing the body at between 900 and 1000°C (1652 to 1832°F), then applying the glaze and firing a second time at about 1350 to 1400°C (2562 to 2752°F).

The hard paste tradition came to Europe in the late 17th century by the sea trade route from China. In the 18th century porcelain ware was made in hongs (factories) in Canton for export to Europe, and the Chinese continued to export ware such as *famille verte*, *noire*, *jaune*, and *rose* well into the late 19th century.(These are French terms referring to 'families' or types with green, black, yellow and pink glaze.)

The first European to succeed in imitating this imported 'china' was an alchemist, Johann Friedrich Böttger, under the direction and help of a Saxony court councillor, Count von Tschirnhaus, and the patronage of Augustus the Strong, King of Poland and Elector of Saxony. Böttger produced hard paste porcelain between 1710 and 1714 in the castle at Meissen, which was the beginning of the Meissen porcelain industry.

With other royal patronage and help from Meissen defectors, a Dutchman, du Paquier, started a factory in Vienna, and from here the secrets of the process

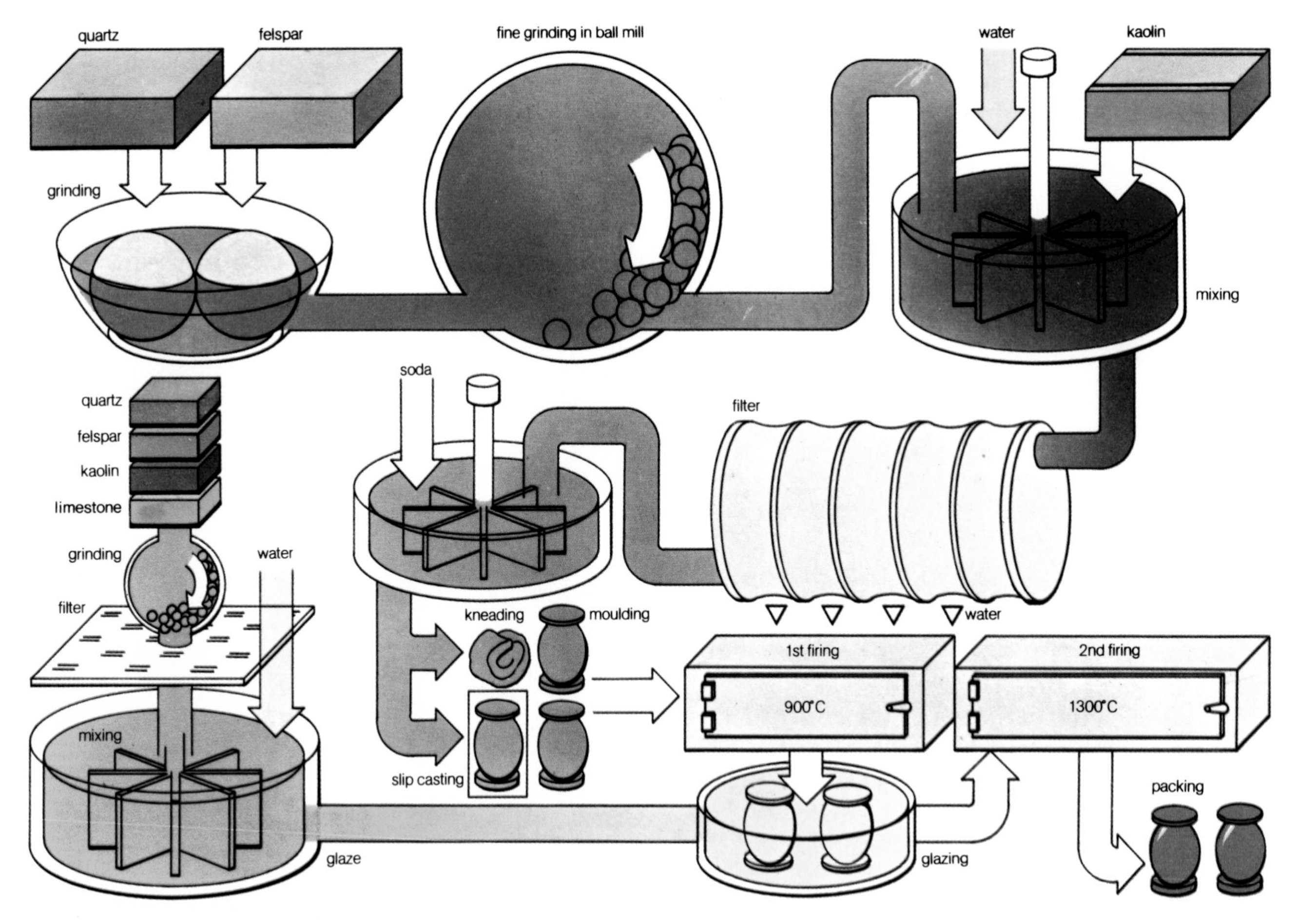

Above: modern porcelain is mostly made by a highly mechanised process, though traditional hand methods and small kilns are still used by craftsman potters. This sequence shows the making and glazing of a typical mass-produced vase. The raw ingredients for the body are quartz, felspar (both hard minerals) and kaolin (china clay). First the quartz and felspar are ground to a fine powder in a ball mill, a rotating drum full of heavy balls which have a crushing action. The powder is then mixed with kaolin and water and, once all the ingredients are thoroughly combined, squeezed almost dry in a filter press like that shown on p. 38. To make the clay easier to knead and mould, soda (sodium carbonate) is added and the mixture is thoroughly kneaded. It is then either moulded or slip cast to shape; the coffee pot on p. 42 is an example of porcelain slip casting. The first firing hardens the article so that it can be glazed. The glaze is prepared from quartz, felspar, kaolin and limestone, and is applied to the article in a liquid form. During the second firing, the enormously high temperature causes the glaze to fuse into a glossy coating, and at the same time to bond closely with the body of the article. This is what gives porcelian its characteristic strength and hardness; in ordinary china the glaze is just a coating.

were disseminated through Europe.

In England, a Quaker apothecary called William Cookworthy (1705–1780) discovered the necessary ingredients for a hard paste porcelain in Cornwall. He held a patent dated from 17 March 1768, and produced ware at Plymouth from 1768 to 1770. The works transferred to Bristol under Richard Champion where it worked from 1770 to 1780, when the formula was bought by a group of Staffordshire potters working at New Hall from 1781 to 1812. The style and body of New Hall was in its turn imitated by Coalport and Caughley.

The manufacturers of porcelain in England were middle class entrepreneurs who had no state or royal subsidies, and so the costs and difficulties of manufacture forced development towards the soft paste types of porcelain, to the total exclusion of hard paste porcelain in England.

Soft paste porcelain manufacturing spread along the overland trade routes from China through the Islamic empire, where an individual tradition developed between the 9th and 13th centuries. These traditions and skills eventually spread from north Africa to Spain and Italy.

The first successful European experiments in soft paste porcelain were done in Florence by Buontalenti at the instigation of the Grand Duke Francesco de' Medici. He used a white burning clay with the addition of a frit (ground glass), which gives a vitreous body when fired at 1100 to 1200°C (2012 to 2192°F). This biscuit ware was then dipped in a lower firing lead glaze, which matures in the second firing at 1050°C (1922°F). Variations on this soft paste theme were tried throughout Europe for 350 years with varying degrees of success.

In England, William Duesbury and Andrew Planché used soft paste at their Derby factory in 1750, and the use of new ingredients gave the wares of different factories their individual characteristics. The Worcester factory (1751–1820) added soapstone (steatite), and bone ash (calcined ox bones) was used by Thomas Frye at Bow (1747–1776) followed by the Chelsea and Lowestoft factories. The next development was the addition of bone ash to the traditional hard paste body by Josiah Spode in 1794, using 50% bone ash, 25% kaolin, and 25% Cornish stone (quartz and felspar). This formula was soon adopted by Derby, Minton, and Coalport, and became the typical English 'bone china' body, an exclusive blend of hard and soft paste porcelain. The porcelain industry in the USA began at the end of the 18th century, making mainly soft paste and bone china. The manufacture of modern porcelain is essentially the same as that for other china ware.

CLOTHING manufacture

The clothing industry is one of the few remaining trades which relies on the basic skills of the operative on the factory floor for the major contribution to the article produced. In the past few years, however, loss of skilled personnel, together with the higher cost of labour and the need for high production rates, has led to the increasing use of automatic and semi-automatic machinery in clothing manufacture.

One feature of the clothing industry which is being superseded by modern methods is the use of outworkers, people who work in their own homes, doing such jobs as stitching or sewing on buttons. The unfinished garments are delivered to them and collected again when the work has been completed, payment being made on a piecework basis.

CUTTING The first stage in the manufacture of all types of clothing is the laying up and cutting of the material. This must be carefully planned so that there is a minimum of wasted cloth, and in the past it required a number of fairly skilled personnel and was done by hand. Today, when up to 100 thicknesses of material may be cut at one time by automatic machinery, the laying up of the material prior to the cutting must be done very accurately, and modern laying up machines are so precise that even checked and striped fabrics can be layed up.

When the material has been layed up on the cutting table, the patterns are marked out on the top layer as a guide for the cutting machine operator. There are now automatic cutting machines available such as the German 'Kuris' system, which accurately cuts lays of fabric from underneath. This machine alone can reduce the number of personnel in the cutting room by some 30%.

THE MACHINE ROOM Many clothing factories now use an assembly line method set up around a conveyer belt system. The conveyer is worked from a control point by one operator who directs work to each machine operator and collects it when it is finished, redirecting it to the next operation in the sytem. This obviates the need for service operators, who carried the work from one machinist to another, and who at one time accounted for 20% of the factory floor labour force.

At one time the only machine used in making clothes was the ordinary industrial sewing machine, call the *flat* machine. These are still in use, but the modern versions incorporate several refinements. Some of the machines made by Singer and Pfaff, for example, have under-bed trimming devices to cut away the threads

Below: a band knife cutter, which cuts many layers of fabric at once. Note the thickness of the pile of fabric and the large heads of the long, slim pins which hold the layers together while they are being cut.

Right: hand work still plays a large part in clothing manufacture, as can be seen from these rows of machinists in a factory in South America. This labour-intensive industry naturally tends to flourish in countries where labour costs are relatively low.

Below left: at the same factory, fabric is washed and stretched to ensure that it lies straight before it is cut out. These rollers are used to stretch the material evenly so that all the threads are straight and parallel.

Below right: at a factory in Peking, knitted cotton socks are being moulded to their final shape by stretching them over heated metal forms.

Bottom: Hong Kong is one of the most important centres of the clothing industry, and is well known for its 'instant' made to measure suits, some of which are being hand finished here.

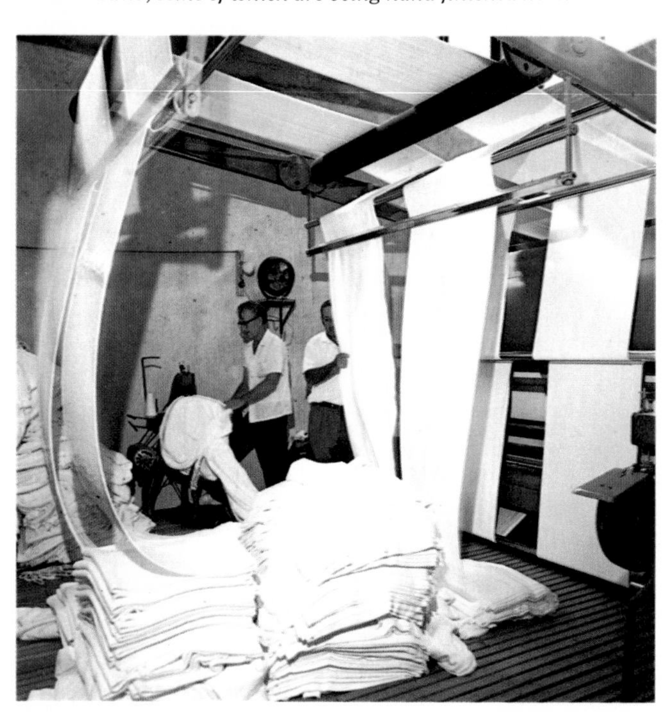

left at the end of a stitched seam. Previously the operators had to trim the seams by hand, but with these machines they need simply to touch a button and the cotton is cut cleanly and the cut ends carried away by a stream of compressed air. Although these machines are expensive, the saving in handling time is enormous and has fully justified their use.

One of the more skilled operations in the making of dresses, skirts and trousers is the insertion of the zip. The need for this skill has been reduced by the auto-

matic zip machines which work from a continuous roll of zip. The operator simply feeds the two pieces of fabric and the zip into the machines which both inserts and makes the zip at the same time. This machine makes a perfect channel zip without the need for a skilled operator, at the rate of 60 zips per hour, far faster than any machinist could achieve on an ordinary flat machine.

Other operations which were considered highly skilled jobs have been turned into simple tasks by the use of machinery. For example, the shaping of collars, cuffs and pockets has been greatly simplified. A simple conversion can be made to any modern flat machine so that by using a shaped plastic tracker to guide the fabric around the needle the operator can produce a perfect shape every time and at a high speed. Collars for mens' shirts can be produced in large numbers by the use of templates which punch out the same collar shape every time. This is only practical away from the fashion side of the clothing industry, where shapes do not change quickly and long runs of one style can be anticipated.

SEAM STITCHING The stitching together of two edges of uneven contour, such as the waisted side seam on a dress, is another process which needed a high degree of skill from the machinist. Today this can be done by an unskilled operator using, for example, the Durhopp Long Seamer, which does the whole operation automatically. This machine is expensive, and relies on long production runs to make it an economic proposition, but where this is possible its high output makes it well worth the investment.

Perhaps the most difficult operation of all in the making of clothing is stitching on the sleeves, which requires much skill in positioning and sewing the material. This can now be done on a machine such as the Pfaff 3801-1/01 sewing unit, which can be programmed to insert the sleeve and stitch tape around the armhole.

PRESSING The final operation in clothing manufacture is the pressing of the finished garment, either by hand iron or mechanical press. In the past the presser spent most of his time removing unwanted creases, but today all creases can be removed before pressing by passing the garment through a steam tunnel. The presser puts in only the creases required.

MATERIALS Until the 20th century all clothing was made from natural materials like wool, cotton, silk, or animal skins and furs. Nowadays, synthetic fibres such as rayon, nylon, polyester such as Terylene or Dacron, and acrylic are used widely in the clothing industry, either alone or in combination with wool or cotton. Some plastics, PVC for example, are also used for garment manufacture, often with the seams welded together by a heat process instead of being stitched. Synthetic fibres now account for over 20% of the world's textile consumption.

KNITTING MACHINERY The earliest examples of knitting were found on the site of the town of Antinoë in Egypt. The world's first knitting machine was designed and built by the Reverend William Lee of Calverton, England, in 1589 and his discovery led to the foundation of the cottage knitting industry which prevailed until the start of this century. After the industrial revolution, power was applied to knitting machines and circular machines were built, which knitted tubular fabric with a constant rotating motion.

Top: a 1923 circular latch needle knitting machine used to make seamless hosiery.

Above: a modern, electronically controlled version making double jersey fabrics.

The two types of machine needle in use are called bearded and latch needles. The bearded needle was invented by Lee in 1589 but the latch was not developed until 1849 by Matthew Townsend of Leicester, England. The latch needle has a pivoted latch element and makes its loop with a simple up and down movement; it is the common needle used on domestic machines. The bearded needle has a springy elongated hook called a beard which has to be closed on each operation of the needle. Although its operation is more complex than that of the latch needle, the bearded needle is still widely used in machine knitting because it can be made in fine gauges and it can be used to knit garment sections 'fashioned' to shape on the knitting machine. In bearded needle knitting the yarn is linked into loops as a first step and as the needle moves down a presser

Above: a flat knitting machine.

closes the beard so that the new loop is pulled through the old loop, which is cast off.

In latch needle knitting the action is continuous and as the needles move down, the hooks take the yarn and pull loops through the old loops. The deeper the needle moves the greater the length of stitch. The gauge of knitting machines is the number of latch needles in one inch or, in the case of bearded needles, the number in $1\frac{1}{2}$ inches.

Circular knitting machines are high speed production machines and can knit rolls of fabric for cutting and sewing in single jersey, double jersey or purl stitch according to the type of knitting machine. Single jersey is knitted when there is only one set of needles, usually positioned vertically round the machine and each in its groove. Wedge shaped cams move the needles up and down and this movement can be controlled for patterning so that those lifted knit and those left down miss the yarn and hold their loops; this is called floating. Colour patterns are made using this system. Alternatively, needles may missknit but still take the new yarn: this is called tucking, and textured surface fabrics are made using this stitch.

Double jersey knitting requires two sets or beds of needles; in addition to the vertical needles in the machine there are horizontal needles, and these pull two faces to the fabric, hence the term double jersey. Purl knitting requires needles with hooks at each end since some loops are knitted to the back and some to the front; the double ended needles require a control

Below: the stitches used in latch needle, bearded needle, purl and double jersey knitting; the miss, used for colour patterns, the tuck, used for textured surfaces; and simple warp knitting.

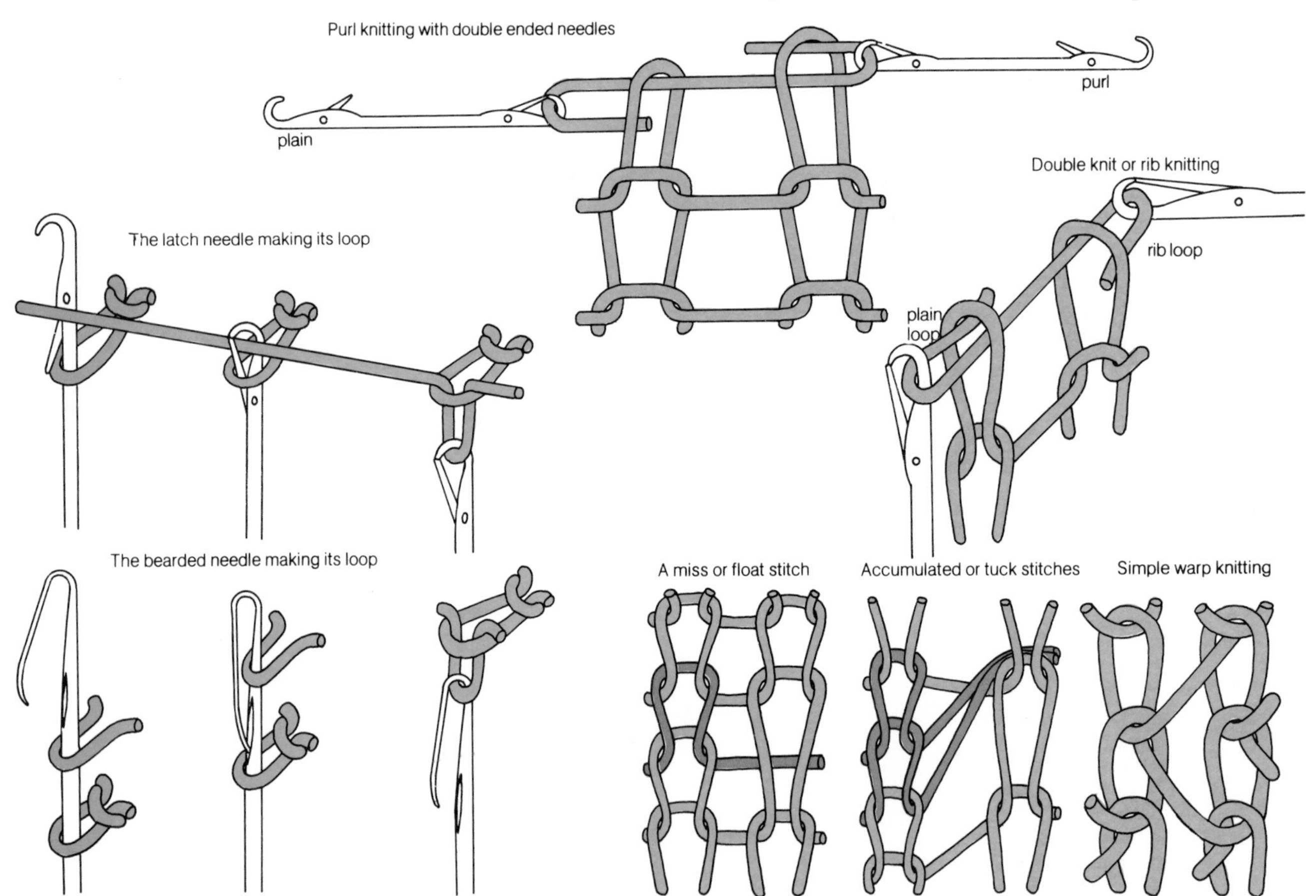

Left: two views of a latch needle flat knitting machine, showing the yarn being fed to the needles and the entire machine, with an endless belt of punched cards controlling its operation.

element at each end.

Combinations of needle positioning and selection for floating and tucking can produce some of the well known knitting structures such as interlock, which consists of two rib fabrics locked together, bourrelet, a double knit fabric where some needles in one bed knit more loops than those in the other to form ripples or waves, milano rib and ponte roma, double knit structures where double knit rows of loops are interspersed with single knit rows, and double pique in which rows of rib stitches are followed by rows of selected single jersey knitting.

The knitting action of a simple flat machine is like double jersey circular knitting except that the needles are in an inverted V configuration. Latch needle flat knitting machines are slower than circular machines but because the knitting is carried out with a to-and-fro motion the fabric has a firm edge on each side called a selvedge. Machines with bearded needles are called 'fully fashioned machines' and are capable of manipulating the loops at each selvedge so that the different garment sections are knitted to shape without any need for cutting afterwards. Flat machines are extremely versatile and the industrial machine is much like the domestic flat machine, but of course built very substantially with a power unit for automatic operation.

The final group covers machines which knit the threads vertically, and this is called warp knitting after the terms used in the weaving industry, where weft describes the threads across the fabric and warp describes those running along the fabric. In one machine for producing shirting fabric, two yarns are fed to every needle in the machine and since the machine has 2500 needles this means that 5000 ends of yarn are fed to the needles simultaneously. Machines of this type can knit over 1200 rows of stitches per minute, giving over three million loops per minute.

STOCKINGS AND TIGHTS MANUFACTURE

Lee's knitting machine of 1589 was quickly adapted to produce fine silk stockings. By the early eighteenth

Left: in the 18th century stockings were worn by men as well as women and were made from silk, wool or cotton, depending on the quality. They were often hand knitted, but the invention of the stocking frame allowed faster rates of production. This was a bearded needle flat machine; the fabric produced was straight and had to be closed by a seam to form a stocking. Here the knitter sits at the frame making silk stockings, while his wife winds hanks of silk thread on to the bobbins used in the frame. She would have spun it herself on the wheel at the back.

Left: modern circular machines make single legs of tights (even more modern ones make one piece tights). They are white; dyeing happens later.

Above: the knitting head of an 8 feed machine. The latch needles are at various stages of their cycles, and the completed tube of 20 denier tight leg can be seen going down the middle.

century hosiery manufacture was an established industry. Labour and trade disputes, however, forced London hosiers to move to the Midlands where over 3000 stocking frames were used in a cottage type industry. Some of these were as fine as present day machines, giving 700 stitches to the square inch.

Circular stocking manufacture began during the last century but did not reach its present production levels until the last 20 years, following the development of stretch nylon yarn.

So widespread is the use of the synthetic fibre nylon that the term 'nylons' has come to be synonymous with stockings and tights. Although other types of fibre are used occasionally, such as silk, wool or cotton, the bulk of the industry uses nylon yarn which has been specially processed by heat setting and twisting to give good stretch and recovery properties enabling the manufacture of stockings and tights which will fit not only just one but a range of sizes. The yarn is delivered to the hosiery manufacturer where it is stored in temperature and humidity controlled warehouses; batch sampling of the yarn is carried out by the laboratory to ensure that correct standards are maintained.

At this stage the yarn is white and is wound on to plastic coated tubes called pirns. The weight or thickness of the yarn is measured in deniers. A denier is equivalent to the weight in grammes of 9000 metres of yarn, and hosiery is made in 15, 20, 30 and 60 denier nylon, 15 being the sheerest.

The pirns are placed on frames above the knitting machines and the yarn is fed into the machine immediately below. Individual pirns are connected by a method called 'double tailing'—when one reel is exhausted a new full reel is automatically brought into operation. About four miles (6 km) of yarn is used for each pair of tights.

The latch needle, invented in the mid 1800s, is a common feature of all hosiery machines used today for the production of tights and stockings. It comprises a hook, at the top, which is opened or closed by a hinged latch, and at the other end a butt or tail, which is fixed to a mechanism in the machine for raising and lowering the needles. There are about 400 to 430 needles in vertical slots in the outside of a cylinder in the machine. These slots extend far below the needles to provide space for the jacks, or patterning devices, which are used when making stitches such as micromesh and run-resist. During knitting the fabric falls down inside the cylinder and is pulled down by a vacuum device, dropping automatically on completion into a bag at the side of the machine.

The three basic patterns used for hosiery are plain knit, micromesh and run-resist. In plain knit the yarn is fed into the open hook which is lowered in its slot until the hinged latch comes up against a previously formed loop which acts as a barrier; this closes the latch as the needle continues to move down and the new yarn is pulled through the old loop which passes over the needle head as it continues to move down (knock-over); having drawn the new loop through the previous one the needle rises again ready to start a new cycle. Shape is introduced by increasing or decreasing the size of the loops.

When producing run-resist or micromesh patterns the needles are raised by the jacks which are in the same slots as the needles and directly below them. Patterns such as these are made by forming two or more loops together. This is done by taking the needle back only to a certain height at which the yarn does not go around the needle stem so that at the next cycle when the needle descends with a new yarn in the hook it collects the previous yarn and both are slipped off the end of the latch together when it is raised to its full height. In run-resist patterns there is a larger number of tuck stitches and every other row is knitted tighter than usual to strangulate the yarn, thus resisting the spillage of loops known as a ladder.

For stockings the machine can knit the entire garment; tights may be formed as extra long stockings which are cut and seamed together to form tights. Machines are now available, however, for producing one piece tights beginning at one toe and proceeding to the first leg; next the body, complete with the waistband; then the second leg and finally finishing with the second toe.

Fully fashioned garments, however, are knitted in the flat form and joined afterwards by a seam up the back. They are shaped by decreasing the number of stitches over the calf.

A modern machine will take about three to six minutes to knit a pair of conventional tights and a one-piece machine about three minutes—a marked improvement on the two weeks which it took a hand knitter to complete a single silk stocking in the sixteenth century.

Products which have not been knitted with a closed seam toe are now delivered to the toe seaming section where the open toe is seamed. The next operation involves straightening as the product coming off the knitting machine naturally tends to be crumpled. Straightening may be carried out on a heated leg form or on a cold wire frame, resembling a large hair pin.

The two single legs are now ready to be joined. The operator feeds two halves into an overlocking machine which first cuts the fabric with a knife and then sews it together. Garments may be joined either without a gusset, forming a simple 'U' seam, or with a variety of gussets, and even full back panels may be used for the larger sizes.

Up to this point the garments are all white and are now ready to be despatched to the dyehouse where they are packed in large sleeves, resembling 'sausages', so that they retain their shape and the dyeing is uniform. During the dyeing cycle samples are taken to ensure that the garments match the master shades.

The dyed garments are then placed on Perspex [Plexiglas] or wire forms for a final quality control check to ensure that there are no flaws. Finally the garments are neatly folded and inserted into individual packets or boxes.

WATERPROOFING FABRICS For fabrics exposed to the elements, water repellency probably represents the most important single property. For example, fabrics which otherwise exhibit a high level of thermal insulation are rendered virtually useless if they are allowed to become wet.

The efficiency of a water repellent fabric is measured by two factors, the quantity of water that can pass through the fabric (penetration) and the quantity of water soaked up by the fabric (absorption). Essentially the penetration depends on the construction of the fabric and the ability of the constituent fibres to shed water. An open fabric will allow water to pass through the interstices, or spaces between the fibre, in spite of a high level of water shedding by the fibres. A closely woven fabric on the other hand will prevent penetration if the water shedding properties are adequate.

Fibres can be divided into two main classes, those that readily absorb water and those which, while not water repellent, have a low level of water absorption.

For comfort in wear, a fabric must be sufficiently

Left: side pattern hosiery being smoothed and formed to its final shape by steam heat. The process resembles that on p. 46, but a lower temperature has to be used for nylon.

porous to allow moist air to circulate while preventing the passage of liquid water. The object is to coat the individual fibres with a continuous film of water repellent substance without binding the fibres to each other or without filming over the interstices. The fibre coating must be tough, yet pliable and elastic, and must adhere tenaciously to the fibre.

Cellulosic (cotton, rayon and linen) and protein (wool and silk) fibres fall into the first class and synthetic (nylon, acrylic, polyester) fall into the second. Fibres of the first class tend to be able to absorb chemicals into the fibre structure more readily than fibres of the second type.

Extremely fine films are required for finishing synthetic fibres. Coarse films tend to exhibit chalkiness (scratching of the surface or creasing of the fabric, causing white surface marks on the fabric).

TYPES OF AGENT Finishing agents for the production of water repellent effects can be divided into a number of basic categories, the choice depending on the type of fibre to be treated, the durability to washing and dry cleaning, and the cost.

The simplest type is based on properties exhibited by untreated natural fibres. Wool, in its natural state, is heavily impregnated with body grease from the sheep, for example, and early water repellent finishes on wool were achieved by treatment with lanolin emulsions. The main objections to this type of finish are the facts that the fabric tends to have an unpleasant greasy handle and the finish is not durable to either washing or dry cleaning. Similarly cotton, which contains wax in the natural state, can be treated with an emulsion of paraffin wax. Fabrics used in the production of tents and tarpaulins are treated in this manner. Durability to cleaning is low, but so is the cost.

Experience showed that it was possible to form a chemical complex between paraffin emulsions and salts of certain metals, particularly aluminium or zirconium, and obtain a finish that showed a higher resistance to washing. Finishes of this type are generally applied to natural fibres and are used on woollen rainwear, high quality tent fabrics and cheaper cotton clothing fabrics.

Extremely high levels of durability to washing techniques such as boiling can only be achieved by means of more advanced chemistry. Again, wax emulsions are used to provide the basis of the water repellent effect, but complex structures are formed by means of pyridinium or melamine complexes. In both cases there is a chemical linkage formed with the cotton fibres.

Two standard tests for proofed fabric: (above left) the British Standard test for surface wetting, set up for demonstration with coloured water and fabric treated on one half only; (below) the Bundesmann test, which measures permeability. The cross-shaped arms revolve under the fabric.

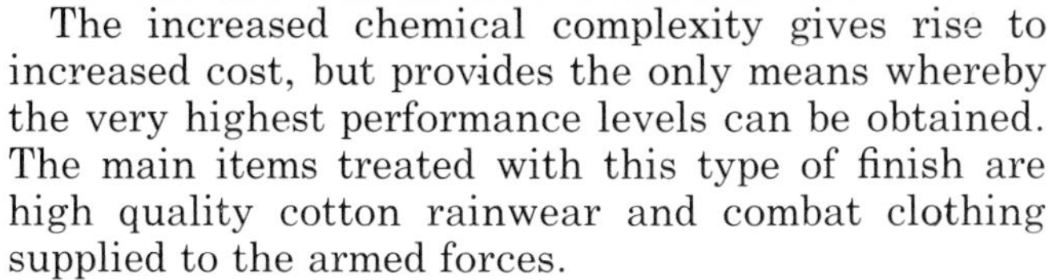

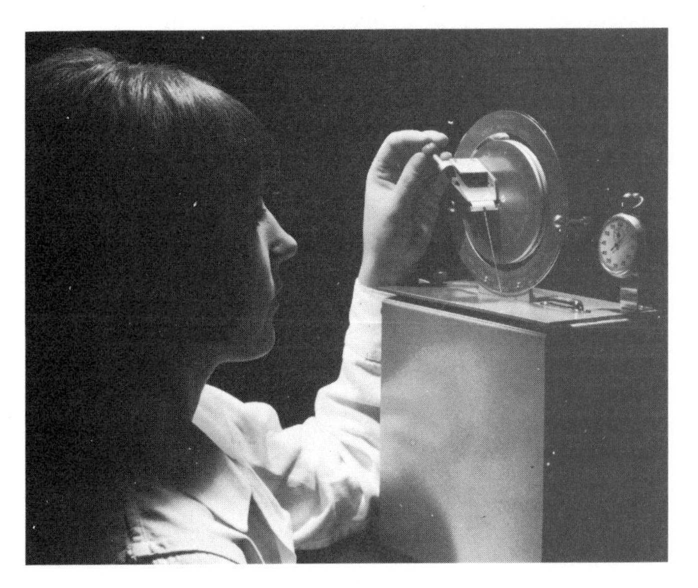

The increased chemical complexity gives rise to increased cost, but provides the only means whereby the very highest performance levels can be obtained. The main items treated with this type of finish are high quality cotton rainwear and combat clothing supplied to the armed forces.

The wax based materials can in general be applied to natural fibres, particularly cellulosic fibres such as cotton or linen, and while varying degrees of wash fastness can be achieved, all have a low durability to dry cleaning.

Wax based films give rise to chalking on synthetic fibres, and in general these fibres are treated with either methyl siloxanes or hydrogen methyl siloxanes (silicones). The films formed are very much finer than those formed by wax and so the problem does not arise. The water repellency effect is achieved by the surface of the repellent rather than by the thickness. As silicones are comparatively expensive, however, their use has been largely limited to synthetic fibres. Silicones exhibit a very high level of durability to washing and dry cleaning and are used on a very wide range of fabrics from rainwear, umbrellas and anoraks [parkas] to furnishings.

As the property of water repellency is based on a surface effect, surface contamination will readily impair any effect and so it is absolutely necessary, for example, to remove all traces of washing agent (such as detergent) if the effect is not to be impaired.

CREASE RESISTANCE Crease resistant, anti-crease, crease shedding, self smoothing, and wrinkle-free are all terms which are used to describe fabrics which show a good recovery from being crumpled or creased. The term crease resistant is a misnomer because virtually all textiles must be capable of being folded or creased, but it is understood at consumer level to mean crease shedding.

The above terms refer to creasing in the dry state, that is, during wear. During the last thirty years wet

Top left: another permeability test. This is the British Standard hydrostatic test. The fabric is not scraped as in the Bundesmann test, but there is water underneath it and pressure is gradually increased from below until three drops have appeared on the surface.

Top: fabrics are treated to give them their crease shedding properties before they are cut up and made into clothing and furniture coverings. The first stage is the impregnation of the fabric with a resin compound, chosen according to the type of fabric to be treated and the use for which it is intended. Dimethylol urea was the first to be used, and this is still in wide use today for impregnating many types of fabric. After impregnation the fabric is dried and 'baked', using the equipment shown here.

Above: apparatus used in a standard test to measure a fabric's recovery from creasing before it is made up into finished items. Speed of recovery is also measured; note the stopwatch on the right.

creasing has become equally important with certain fabrics. Easy care, non-iron, wash and wear are terms which have been used to imply that individual textile articles, if laundered as recommended, would be sufficiently crease free to enable them to be used with little or no ironing or pressing. The situation is further complicated with the introduction of the term 'durable press' to describe garments which will retain their shape and appearance during washing and wearing, and this usually includes a durable or permanent crease in trousers. It is now possible to have good dry and wet crease recovery together with good crease retention, necessary say in the case of permanent pleats, in the same fabric and garment.

The tendency of fabrics to crease depends on the fabric and yarn structure as well as on the properties of the fibres from which they are formed. If the yarns and fibres in a fabric are relatively free to move, as in a knitted or loose woven fabric, the tendency to retain creases is less than in a tightly woven fabric. Ultimately the creasing behaviour of a fabric depends on the fibre properties, particularly fibre extensibility (ability to stretch without breaking) and recovery from extension under the creasing conditions. Fabrics made from rayon and untreated cellulosic fibres, such as cotton and flax, have a relatively high tendency to crease, and since 1920 scientists have been working on methods of producing crease shedding cellulosic fabrics, usually by applying thermosetting resins (types of plastic which are heat resistant), and fibre crosslinking agents to the fabric.

From about 1950 the synthetic fibres, particularly nylons (polyamides) and polyesters, have introduced new standards not only of crease recovery after wear, but also in easy care properties after laundering. These, and most other synthetic fibres are thermoplastic, that is, soften under the action of heat, and their crease shedding properties can disappear if creasing is carried out at a sufficiently high temperature, which means that some care has to be taken in laundering. It also means that deliberately inserted creases can be retained and this property has been used extensively in putting durable creases and permanent pleats into polyester blend trousers and skirts respectively.

The standard method of imparting 'crease resistance' to cellulosic (cotton, viscose rayon and cellulose acetate rayon) fabrics consists of impregnating with the resin formulation, drying, curing by baking or possibly steaming at a high temperature to polymerize the resin and crosslink it with the cellulose molecules. The fabric is then washed to remove any surplus chemicals. The resin formulation consists essentially of unpolymerized or partially polymerized resin and a polymerization catalyst, for example ammonium dihydrogen phosphate or magnesium chloride. The first resin to be used successfully, and still widely used today, was dimethylol urea, formed by reacting two molecules of formaldehyde with one molecule of urea. Often this simple resin is made in the dyehouse, but also today there are many proprietary formulations using dimethylol urea or more complicated resins such as melamine-formaldehyde, dimethylol ethylene urea (DMEU), dihydroxy-dimethylol ethyleneurea, and dimethylol propylene urea (DMPU).

Dimethylol urea gives a satisfactory finish for many types of fabric used for dresses, blouses, casual shirts and furnishings. The newer types of resin, of which dihydroxy dimethylol ethylene urea is the most important, are used to provide finishes which will withstand high temperature washing. They are also used in 'permanent press' or 'durable press' finishing in which the resin may be crosslinked after the garment has been made. This gives the garment a durable shape and appearance retention.

FIREPROOFING FABRICS The term 'flame retardant' is used to describe fabrics which will not support combustion and are self-extinguishing. Fabrics of this type, when involved in an accidental fire, would not contribute to spreading the flames. Other descriptions, such as flame proof, fire proof and flame resistant, are

Near left: a polyamide fabric first melts, then burns; burning drops can cause severe injury.

Centre: the garment on the left has been treated with Proban, a flame resistant finish, and laundered fifty times. It is only charred by a flame, while the untreated garment on the right is completely consumed within minutes.

Right: testing a textile with a radiant panel.

either meaningless or misleading.

Nearly all fabrics are combustible to some degree. The rate of burning ranges from that of guncotton (nitrocellulose), which burns so quickly that it produces an explosion, to that of asbestos, which is virtually unaffected by fire.

Flame retardant fabrics can be produced in two ways: by making them of fibres which do not burn, or by chemically treating the fibres to produce the desired effect.

Asbestos, steel and glass fibres are non-combustibles, but the yarns produced are suitable only for certain applications. Synthetic fibres designed for their flame retardant properties are now being manufactured from polyvinyl chloride (PVC), from a copolymer of acrylonitryl and polyvinylidene chloride and from aromatic polyamide.

Fabrics produced from these synthetic fibres are similar in textile characteristics to polyamide (nylon) and polyester fabrics. They have several disadvantages however. They are difficult to colour, requiring special dyes, expensive machinery and higher temperatures than other fabrics. They are difficult to launder, being sensitive to temperatures because of their thermoplastic nature; they cannot be boiled to remove stains, and drying them at too high a temperature may result in permanent wrinkles. Most importantly, although they do not burn, some of them give off poisonous gases when subjected to very high temperatures.

Chemical modification of existing fibres, both natural and man made, can be accomplished by adding chemicals which either change the burning chemistry of the fibre or break down under the influence of heat and form a fire extinguishing product.

Cellulosic fibres such as cotton, linen or viscose rayon represent the class which is most readily subject to changes in burning chemistry. Pyrolysis (chemical decomposition by strong heat) leads to rapid decomposition of the chemical structure at temperatures above 300°C with the evolution of flammable gases and the production of tar-like materials. When the temperature reaches 350°C the flammable gases ignite, a lot of heat is produced, carbon dioxide is given off and the cellulose is reduced to ashes. If cellulose, however, is treated with a powerful dehydrating agent, such as concentrated sulphuric acid, it will decompose at a lower temperature and will not burst into flame. Furthermore, no flammable by-products are produced. Chemicals used to produce this flame retardant effect are mono ammonium phosphate, mixtures of borax and boric acid or dicyandiamide phosphate. The decomposition will occur at temperatures of 250 to 290°C. Finishes of this type do not affect the normal textile characteristics of the fabric but are not durable through cleaning processes.

Fastness to washing and dry cleaning is usually

required of flame retardant fabrics. For this reason the preference is for reactive organo-phosphorus compounds, which can be rendered insoluble after application. These chemicals are used for most flame retarding treatments today. The application can be carried out either at the fibre manufacture stage (for viscose rayon) or as a simple immersion treatment after the weaving or the knitting of the fabric.

Substances containing bromine (such as ammonium bromide) are used to produce flame retardant effects by the fire extinguishing method. Under the influence of heat the chemical decomposes, liberating the dense, heavy gas bromine which smothers the fire by not admitting oxygen. Substances of this type are soluble in water and therefore not fast to laundering. Other chemicals which are not soluble will give the fabric a heavy, sticky feel. In addition, the fire extinguishing properties are rendered useless in a breeze, which blows away the bromine gas blanket.

Fire extinguishing effects can also be achieved on cellulosic fabrics by adding a mixture of antimony oxide and a chlorinated compound such as polyvinyl chloride. When a flame is applied the chlorinated compound decomposes, giving off hydrogen chloride which reacts with the antimony oxide, producing antimony chloride. This compound is a powerful flame suppressant.

COFFEE production

The stimulating effect of the coffee berry was originally discovered in the third century in Abyssinia. According to legend, a shepherd noted that his flock became livelier after eating the fruit of a certain plant, but it was the Arabs during the thirteenth century who first used the roasted seed or bean to make a beverage. The habit was perpetuated by Middle Eastern pilgrims and nomads, and gradually spread to Europe.

The plant flourishes in a warm dry climate at heights between 1500 and 6000 feet (450 to 1800 m), and is indigenous to Abyssinia and Arabia. Until the seventeenth century Arabia was the only source of supply, but during the seventeenth and eighteenth centuries, when coffee houses became fashionable among Europeans, the plant was transplanted to their colonies in the Indies and the Americas. Coffee is now grown successfully between the Tropics of Cancer and Capricorn, and the principal producing countries include Brazil, Colombia, the Ivory Coast,

Below left: this girl, picking coffee in Southern Brazil, may pick 45kg (100lb) a day.

Centre: open air drying of beans in El Salvador. Drying machines at rear may be used if needed.

Right: hand sorting of beans in West Germany; mouldy beans may ruin the taste and aroma.

Kenya and India.

The coffee plant is an evergreen which can grow to a natural height of about 20 feet (6 m) but when cultivated commercially it is pruned to a height of 6 to 8 feet (about 2 to 2.75 m) to encourage a fuller bush and allow easier harvesting.

The two botanical species normally cultivated are the *arabica* and the *robusta*. The latter is considered inferior as it produces a coarser tasting coffee than the finer flavour of *arabica*, which unfortunately tends to diminish during processing. The stimulating effects of coffee beans are caused by the organic compound caffeine.

The ripe fruit resembles a cherry and contains a yellow pulp which surrounds two oval seeds or beans. The beans, which lie with their grooved and flattened faces together, are covered with a fine silver skin and wrapped in a tough outer skin known as the parchment. After harvesting the beans are cured by either a wet or dry process.

CURING The dry method, which is generally favoured in countries where water is in short supply, involves drying the beans either by artificial means or by spreading them on the ground to dry in the sun. Frequent raking is necessary to ensure the beans dry evenly. A hulling machine is then used to remove the parchment and pulp from the cured bean.

When the beans are cured by the wet process, the red pulp is first removed by machine, then the beans, which are still in their protective parchment, are

Below: raking over beans in the Sudan, for even drying during the two to three week process.

soaked in water for up to 72 hours to ferment any pulpy sugar still left. After the water is drawn off, the beans are thoroughly washed until the parchment is quite clean. The beans are dried and the remaining skin is removed by a milling machine. The result of either of these processes is a cured bean which is generally called 'green' coffee.

Coffee is also produced without the stimulating properties of caffeine. The process involves breaking up the green beans by steam and subjecting them to the action of either chloroform or some other solvent to dissolve the caffeine. The coffee is then steamed again to remove residual solvent and dried. After roasting the decaffeinated coffee will taste similar to the usual kind.

When the curing process is completed the coffee is tested, graded and offered for sale in the markets of the world. Supplies of green coffee can be bought either direct from the country of origin or from the warehouse stock of a coffee broker. As the flavour of the bean varies according to type, growing conditions, method of drying and other factors, coffee buyers require a very specialized knowledge in order to choose the right beans for their standard blends.

The international coffee trade is affected by natural, political and economic factors; for example, crop damage in 1953-54 resulted in record prices and extensive new planting. When the fruit of the new plants reached the market in the late 1950s the supply exceeded the demand. In order to allocate supplies and stabilize the market, an International Coffee Agreement was signed in 1962 and an International Coffee Organization set up with headquarters in London.

ROASTING After blending the beans are roasted, a process which requires accurate timing to ensure that the porous cells will transfer their flavour and aroma more readily during the brewing stage. The roasting is done in rotating drums which are usually gas heated, but without coming into direct contact with the heat source. The beans are then released into a large circular perforated metal tray, where they are stirred by revolving metal arms. Rapid cooling helps preserve the aroma of the coffee by closing up the pores of the bean.

Before the beans can be brewed into a beverage, they must be ground so that the full flavour will be extracted. This can be done by the consumer, but is often done by the manufacturer, who then packs the finished product into vacuum sealed jars or tins to preserve the freshness.

INSTANT COFFEE An American chemist invented

Below: instant coffee is made by one of two methods; spray drying or the more expensive freeze drying, which preserves more flavour. Beans are roasted and ground, then brewed with hot water as if making real coffee, though by the most economical process possible. The liquor is then dried by spraying it through a stream of hot air, leaving a powder which is sometimes agglomerated into granules; or foamed, frozen and vacuum treated to drive off the water.

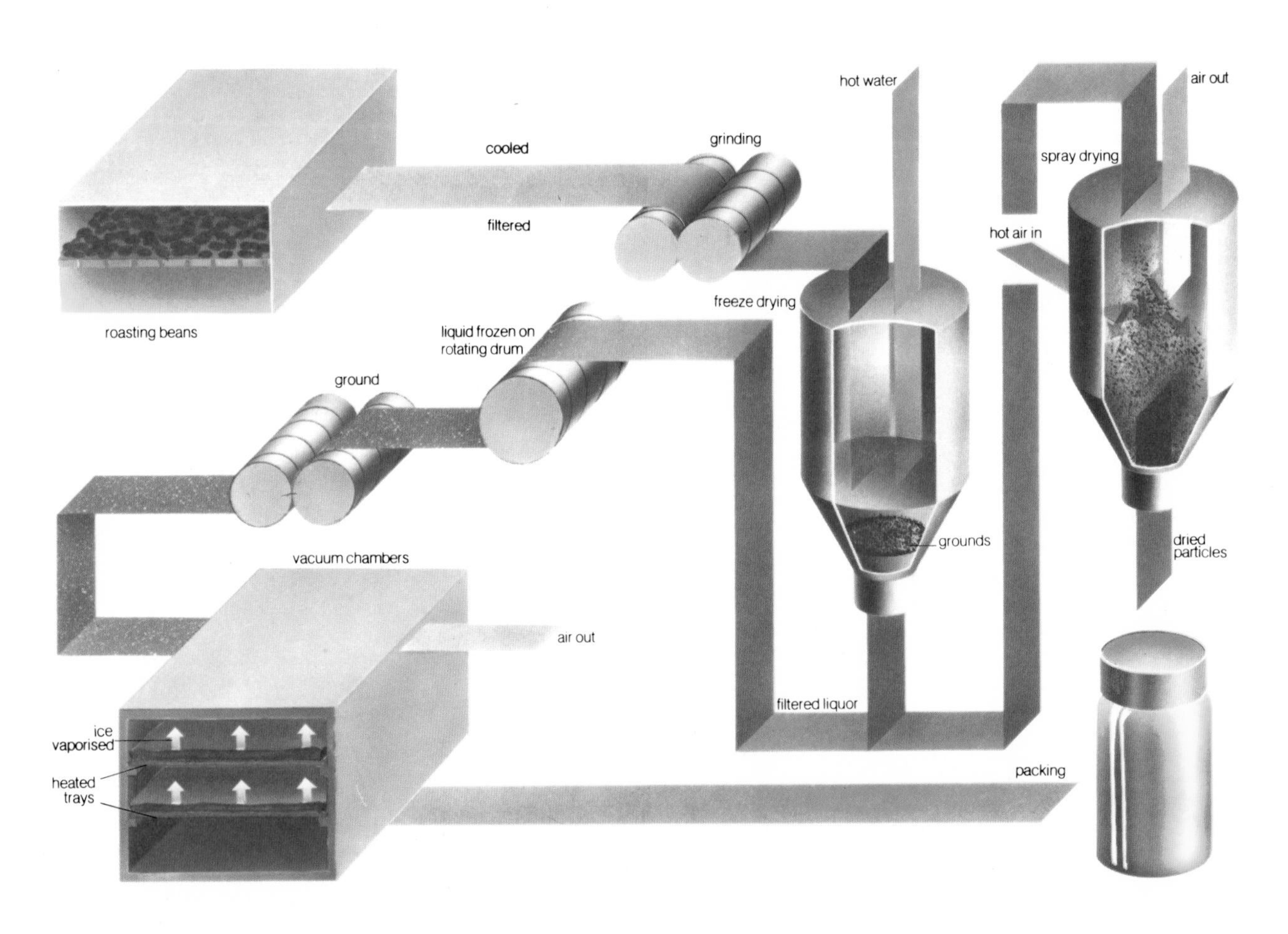

a practical soluble coffee, nowadays called instant coffee, in 1909, but it did not become a commercial success until World War II, when it was widely issued to US troops in the field. After 1946 instant coffee rose in popularity in the USA until it captured more than 25% of the coffee market. In Britain, coffee drinking, which had declined in popularity after the eighteenth century, has revived since World War II, and 90% or more of the coffee consumed in Britain today is instant coffee.

There are three methods of producing instant coffee; the initial stages of roasting, grinding and brewing are similar in all three processes. First the green beans are cleaned and blended. Blending brings together the qualities of different varieties and enables manufacturers to produce a more economical product. Supervision by the blenders and meticulous quality control ensures that required standards are met.

The selected blend of beans is roasted, cooled and filtered before being fed into a granulizer which is adjusted to give the type of grinding required. Suitable particle size is necessary to ensure that the full flavour will be extracted (brewed) from the cells without crushing them. The ground coffee is then conveyed to the extraction plant where hot water is pumped through it. When the required strength is obtained, the brew is pumped through a clarifier and cooler, leaving the spent grounds to be ejected from the extractor. The liquor is then ready to be processed into instant coffee.

SPRAY DRYING The spray drying process is the cheapest and most common method. The coffee brew is injected as a fine spray into a chamber through which a stream of hot air is blown. The water in the droplets evaporates and the vapour is removed, leaving behind the particles of dry coffee which are then conveyed to the packaging machinery. Soluble coffee must be sealed as quickly as possible into moisture proof containers; otherwise it will quickly deteriorate.

In a more recently developed process, after spray drying is completed the product is partially re-wetted by steam or water or both to bind the powder together in large granules. This alters the appearance of it but not the taste.

ACCELERATED FREEZE DRYING Freeze drying is the most expensive method of producing soluble coffee but one which retains more of the coffee flavour. Thus even though it results in a more expensive product, it has meant an increased consumer acceptance of instant coffee, particularly in the USA.

After extraction the liquor is foamed with gas to produce the right bulk density in the finished product. It is then frozen on a belt or drum in the form of a ribbon. While still frozen it is broken, ground and sieved into the desired particle size, and placed on trays which are conveyed into a vacuum chamber and lowered onto heated plates. The combination of applied heat and vacuum conditions causes the coffee to dry without returning to a liquid state. After drying the product is returned to normal conditions of pressure and temperature, and packed.

COINS and minting

The earliest known coins in the Western world were made about 700 BC in western Asia Minor from a naturally occuring alloy of gold and silver known as electrum. Four hundred years before that, however, the Chinese were making 'spade' and 'key' money; the spade was shaped like the digging implement, while the key resembled a modern Yale key.

The Chinese coins bore their denomination, but unlike those of Asia Minor, they were cast from moulds which carried the design. The coins from Asia Minor were stamped on one side with a tool bearing the design, rather than cast; this technique was the true forerunner of modern minting techniques.

MODERN MINTING METHODS The Coinage Acts of the various countries set rigid legal requirements on the composition of the coins as well as the size and weight. For this reason modern minting methods begin with standard non-ferrous industrial techniques.

The alloys used are melted in furnaces and poured into moulds of rectangular cross section. Some modern installations include horizontal continuous and vertical semi-automatic casting plants, which means that the stock can be moulded in continuous lengths and cut off as desired.

Next the stock must be rolled under pressure to make it the right thickness for the particular coin. Stock of thick cross-section is hot rolled, while thinner stock is cold rolled. The rolling process causes work-hardening of the stock, which then requires annealing, a process of heating and cooling which makes the

Above: this 1750 engraving shows round coin blanks being stamped out of sheet on the right; at the front is a die stamping press which impresses the design on one coin at a time. The momentum of the heavy metal balls on the ends of the arms, and the powerful swing imparted to these by the chains, causes the blank to be struck with great force.

stock soft enough for further rolling and cutting.

Once the rolling has been completed the stock, which can be in long lengths or in coils, is passed to machines with multiple cutting tools which cut it into blanks at a speed of up to 400 strokes a minute. Because of the annealing process and the use of lubricants during rolling, the blanks are stained and oxidised on the surface. They must now be pickled in dilute acid, usually in revolving barrels which also allows them to burnish one another, giving a brightness to the chemically treated surfaces. The blanks are then washed clean and dried in hot air.

Some coins are difficult to shape from flat blanks, particularly when there is a wide step or rim around the edge. In this case the blanks are rolled in grooves under pressure to force up the metal around the edge until the edge has a greater thickness than the centre. This process is called rimming, or upsetting. When coins have security designs on the edge the design is rolled into the edge of the blank after rimming.

MINTING TOOLS Pairs of dies are required for the striking of the designs into the blanks, one die each for the head and the tail of the coin (obverse and reverse in minting terminology). The manufacture of the die begins with an artist's model of the design, six to eight times the size of the finished coin. A nickel-plated replica of the design is produced, which is then mounted in a reducing machine. This is a tracing device which machines an exact replica of the design in tool steel to the scale of the coin to be produced. Tool steel is a very high quality steel which

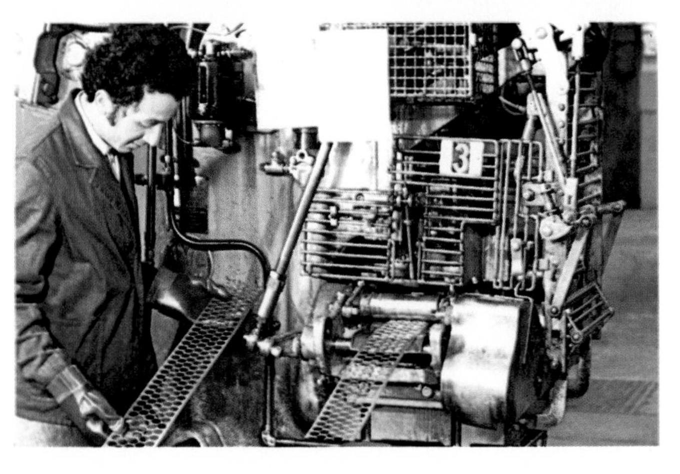

can be hardened by heat treatment, a heating and cooling process similar to annealing except that it makes the metal hard rather than soft. The steel must be heated to a precise, very high temperature, and cooled at a specified rate in a certain type of bath. The exact specifications of the process depend on the type of steel being treated.

From the finished tool, known as the reduction punch, a series of tools is made by hobbing (striking the design into a new die steel). These are the matrix working punch, and working die, one made from the other in succession.

COINING Coining presses are capable of producing coins one at a time or up to four at a time. Those which strike singly can work at a rate of 400 coins a minute where loads of up to 180 tons of pressure are required. Heavier presses capable of 250 tons normally work at 120 coins a minute.

A collar is installed in the press. This is a block of steel which has been bored out to the size of the finished coin. If necessary the collar has grooves cut in it to give the coin a milled edge. The sequence of striking is carried out by the press pushing the blank into the well of the collar, in the bottom of which is installed one of the dies. The other die then travels down from above and, on impact, causes the design to be impressed on both sides of the blank. At the same time the pressure is sufficient to cause the metal blank to spread out to fill the collar.

Surface conditions are very important in coining, so the blanks are usually held to a maximum centre line average which limits surface irregularities. Mints with modern equipment use X-ray fluorescent spectrometers to analyze alloys; otherwise chemical analysis is used.

After coining, the diameters of coins are checked with micrometers or other types of gauges. The weights are checked by measuring the weight of a specified number of coins against a stated weight to which is added an allowable tolerance.

Sometimes coins are accidentally double stamped or have other defects. Such coins rarely get through the inspection process, but when they do they quickly become valuable collectors' items because of their irregularity.

COLOUR PRINT making

Photographic colour prints may be made by a wide variety of processes but the most important is the negative-positive process, which produces a colour print from a colour negative camera film.

Colour film contains three separate layers, each of which is able to record light from approximately one third of the visible spectrum, corresponding to the three primary colours red, green and blue. When the material is processed, the image of the scene in each layer is produced in a dye complementary in colour to the light which produced it. Thus the image in the red recording layer is made cyan (blue-green, white

Left: a reducing machine used in modern die making; it follows the contours of the original oversize model and reproduces them on an actual size, 'right way round' master from which the stamping dies are reproduced in negative.

Below left: a general view of the reducing machine: the round object on the right is the large version of the design.

Bottom left: a blanking press which can produce 400 blanks a minute. The man holds a sheet from which blanks have been struck; this will be melted down and re-used.

Below: the blanks must next be prepared for striking the design. To make the material soft enough to receive the impression of the die, they are annealed by heat treatment, and then the surfaces are cleaned in acid to remove the film of oxide which would mar the finish.

Bottom: finished coins are checked for flaws, surface scratches, discoloration, faulty impressions and so on. Special high-quality coins, of what is known as 'proof' quality, are also made for collectors; ordinary circulated coinage is less than perfect.

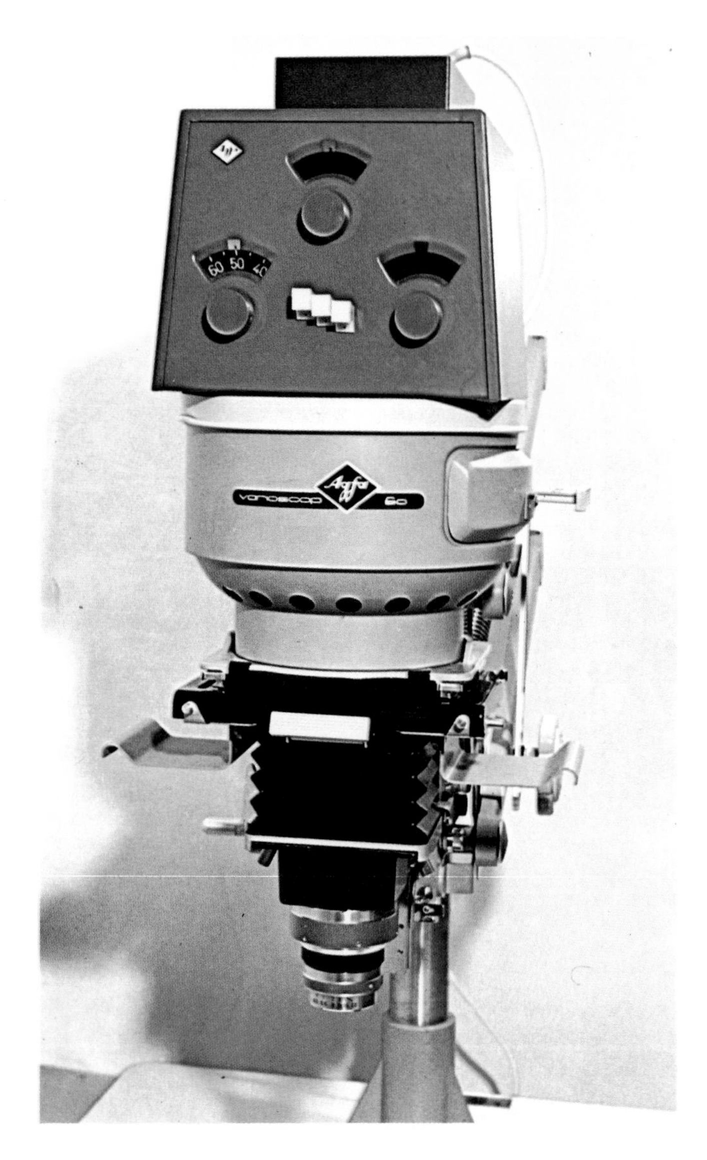

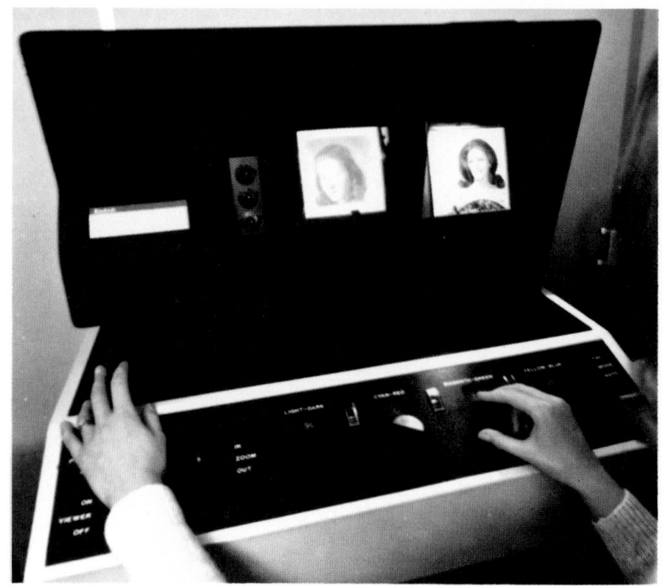

light minus the red component), that in the green recording layer is made magenta (blue-red, white light minus the green component) and that in the blue layer is made yellow (white light minus the blue component). In making a positive colour print from the colour negative the same functions are repeated, using a paper material incorporating the red, green and blue recording layers.

Because the dyes in the colour film are not very pure, a print made in this way would have muddy (desaturated) colours. To overcome this, a mask is incorporated in the film, correcting these deficiencies. This mask gives the film its typical orange colour.

Colour prints are usually made from a colour negative, using an enlarger to project an image of the negative onto the paper. Such an enlarger may be adapted for producing colour prints by making provision for the use of coloured optical filters—light absorbing sheets—in the path. The position chosen

Above: a colour enlarger for making colour prints of any size. The negative is placed in the tray above the bellows; the three dials on the blue painted light box above alter the cyan, magenta and yellow content of the light.

The bottom row shows stages in the production of a colour print. First: the negative has an orange mask and colours complementary to the originals: magenta for green and so on.

Second: a standard colour print analyser would assume equal amounts of each colour and adjust the printing filters accordingly. The result, as too often accepted by the customer, gives a greenish cast to this prevailingly pink picture.

Third: if the customer has complained about the first print, or if the job is being done with proper care in the first place, the print is then remade, adjusting the filters so that green is decreased and magenta increased.

will depend upon the way in which the prints are to be exposed.

The simplest method is to make three successive exposures through a deep red filter, a deep green filter and a deep blue filter. These expose the red, green and blue recording layers of the print material individually. This is known as additive or tricolour printing, and requires only the three filters mentioned, which may be placed conveniently below the lens.

A faster and more often used technique is called white light printing. In this case a single exposure is made to white light which has been adjusted in colour by means of pale yellow, magenta or cyan filters within the enlarger. The filters are adjusted to ensure that each of the three layers is correctly exposed in the same time. The filter pack used to adjust the colour of the exposing light is made by selecting from a series of filters of varying depth of colour: to operate satisfactorily a set of 18 filters is required, 6 in each colour.

The top row of three pictures shows the making of professional colour prints in quantity. First: a Video Colour Negative Analyser: the negative is placed under a scanner on the left and a true colour TV picture of it appears on the centre screen. This is compared for colour with the standard print on its right. Colour and brightness can be varied by four dials, using a zoom control to check a small area if necessary. The data is punched automatically on paper tape which accompanies the negative to the printing head.

Centre: here, the tape ensures that the correct filtering and exposure are given; it can be overridden if necessary.

Right: after loading into a processor (in the dark) and processing, the prints are dried.

Professional colour enlargers often incorporate filter assemblies to allow the colour of the printing light to be altered continuously over a wide range by simply turning a wheel.

The exposed paper must then be processed to produce the colour image. Different manufacturers' materials will require their own particular processing chemicals and procedures. The essentials common to most processes are colour development and bleach-fix.

Colour development produces an ordinary black and white image for each colour together with an associated coloured dye image.

During the bleach-fix stage, the bleaching action removes the silver which produces the black images, leaving only the coloured dye image. The fixing action removes all the silver compounds, leaving the material stable against further exposure to light. In some processes the bleaching and fixing are produced by two successive solutions used in that order. Finally, the print is dried.

There will be one or more intermediate washing stages to remove unwanted chemicals, and possibly additional treatments to improve the final quality.

The printing exposure and filtration requirements of individual colour negatives vary greatly because of manufacturing differences in film and paper, camera exposure conditions, variable processing of film and print, and the colour of the unfiltered enlarger lamp. In general, amateur print makers must operate by trial and error. A first print is made with a range of exposures and an average filter pack. After processing the trial print may be inspected and the correct exposure time determined. The print is likely to have a colour cast—an overall colour bias—due to the use of an incorrect filter pack. If the print has an overall yellow cast, this may be corrected by increasing the amount of yellow filtration, a green cast may be corrected by reducing the amount of magenta filtration and so on. Having made appropriate changes to the filter pack another print is made, which may show the need for further adjustments to filtration and exposure before the final print can be achieved.

This procedure is too wasteful of material and time to be considered for professional or industrial use. For such applications it is usual to measure the amounts of red, green and blue light transmitted by the image on the negative by means of a photoelectric cell and filters. Comparison with the readings on a similar 'master' negative allow the correct filtration to be determined directly. This is made easier if there is a known grey patch, with equal amounts of all colours, on the film. Alternatively a very flexible method, always employed for the mass production of amateur colour prints by automatic printers, uses the principle of integration to grey. One or more photoelectric cells are used to assess the red, green and blue content of each negative and an exposure is given based on the assumption that the picture content is equivalent to an even grey colour. For a large percentage of amateur pictures this procedure produces acceptable prints. When the picture contains a large area of one colour the resulting print will have a cast of the complementary colour and is said to suffer from 'subject failure'. Such prints should be remade with a corrected setting of the printer controls by hand.

OTHER METHODS Colour prints may also be produced directly from colour slides (positive transparencies) using a type of colour paper requiring more complicated processing. Owing to the absence of the colour correction masks, the colour quality is not so good as for the negative-positive process.

For specialized uses other processes are available. In dye-transfer, the print is assembled by successive transfer of dye images from separate records of the red, green and blue content of the picture. Dye-transfer prints are of high quality and allow a great degree of manipulation of the final image. Dye-bleach processes produce pictures by destruction of unwanted dye from a material containing maximum amounts of dye in each layer prior to processing, instead of the more usual production of wanted dye. Dyes of better colour saturation and higher resistance to fading can be used in dye-bleaches.

DIAMOND cutting

Rough natural diamonds can be divided into four principal shape categories: stones, cleavages, maccles and flats. Very few diamonds are flawless. Most contain spots and various other types of inclusion, usually carbon, and these may be invisible to the naked eye. The position of flaws, and their size, is important in the cutting of a clean gemstone.

The principal diamond cutting centres are Antwerp, Bombay, New York, Amsterdam and Tel Aviv, but cutting is also carried out in many other countries. Each centre has grown accustomed to cutting a particular category of diamond; Antwerp specializes mainly in cutting cleavages, maccles and chips, the United States in stones, while Amsterdam, Israel

Bottom left: cleaving a diamond. A mark called a kerf has been scratched on the stone with another diamond. A metal blade is inserted in the kerf and tapped sharply in the direction of the plane of cleavage.

Below: bruting a diamond. It is roughly rounded by pressing another diamond against it while it spins in a chuck.

Bottom: polishing a diamond. Facets are polished by mounting the stone in a device similar to a record player arm which presents it to the polishing wheel. Diamond polishing powder is used. The operator's skill is essential in presenting the stone correctly.

Right: the small picture at the top shows how two brilliant-cut diamonds come from a stone, or full crystal. Cleavages, maccles and flats are less regular fragments.

Next from top: three views of a brilliant, naming the facets; there are 57 or 58, depending on whether there is a flat bottom or culet.

Bottom: the brilliant cut gives the most 'fire' to a diamond because of the way it splits white light into colours. Here the angles are exaggerated for clarity.

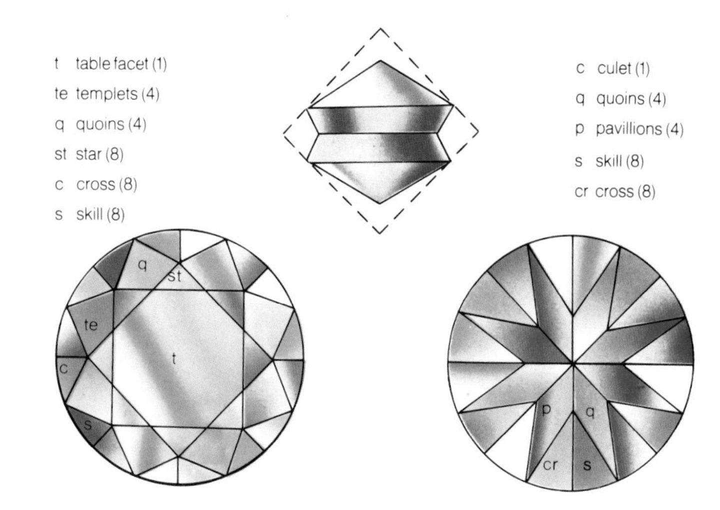

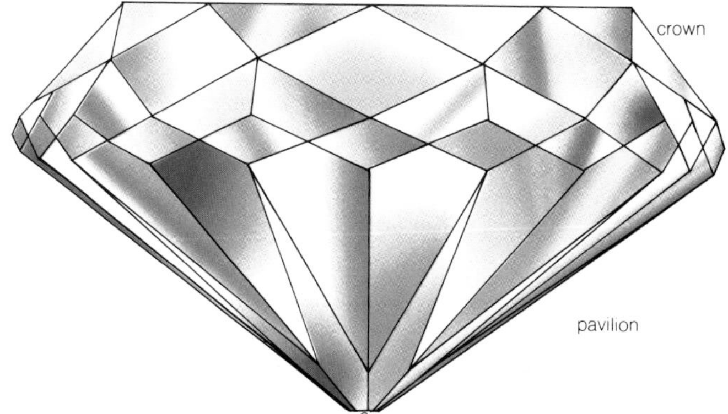

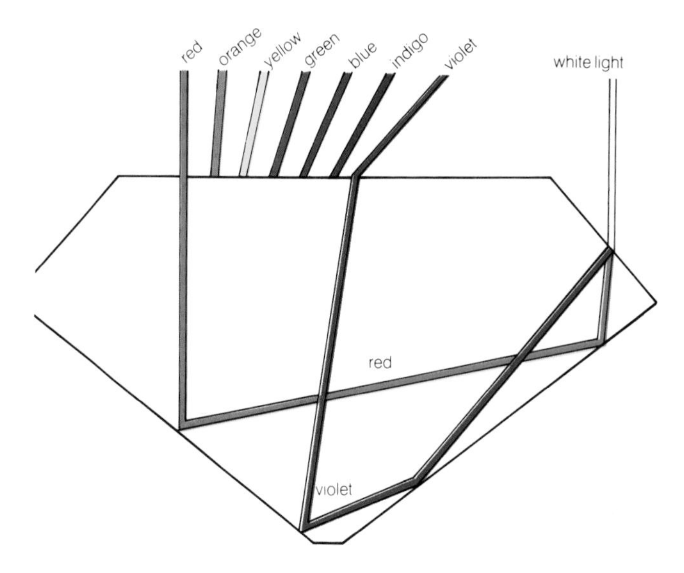

and India cut mostly small brilliants.

Although it is the hardest natural substance, a diamond can be shaped and faceted. This is because it has planes of relative weakness, along which, if expertly handled, it can be cleaved cleanly in two. A diamond crystal has hard and soft faces, and on each face there are hard and soft directions (relatively speaking: even the soft direction is incredibly hard). Therefore, if diamonds are crushed a powder is obtained which can be used to polish other diamonds, since when polishing, many of the particles in the powder present their hard directions to the soft direction of the diamond face being polished.

The four processes in making a gemstone are cleaving, bruting, sawing and polishing.

CLEAVING Cleaving is used before polishing to obtain an improved shape from an irregularly shaped stone, or to split the rare large stones into manageable pieces. A skilled and experienced worker can determine the cleavage plane direction and then, on an edge which is relatively soft, make a scratch mark called a kerf with the hard edge of another diamond. A metal blade, with its plane parallel to the cleavage plane, is then inserted in the kerf, and tapped sharply with a wooden mallet; if this is done correctly the diamond will cleave in two.

SAWING Preliminary shaping is often done with a high speed slitting saw, a thin disc of phosphor-bronze run at about 5000 rpm. The edge of the disc is covered with a paste of fine diamond dust and olive oil. It is a particularly useful process when cutting across a cleavage plane.

BRUTING Bruting is a rounding process carried out before polishing. The diamond to be polished is mounted in a lathe chuck and another diamond pressed against it as it revolves, roughly rounding it.

POLISHING Diamond powder is rubbed on or impregnated into a cast iron wheel known as a scaife. The diamond to be polished is mounted in a dop and tang: the dop is the holder into which the diamond is fixed with solder, or held mechanically. The tang, acting rather like a record player arm, presents the diamond to the revolving scaife. Because of diamond's hard and soft directions the correct presentation of the diamond to the wheel is one of the essential skills of the diamond polisher.

The most widely used cut is the Brilliant cut, which requires 58 separate faceting operations. A large diamond can take months to polish. Other popular cuts are the Emerald, Oval, Pear and Marquise.

Polishing has been traditionally the craft of the individual. Recently, however, a new machine, The Piermatic, has been developed which automatically polishes diamonds up to half a carat with great efficiency.

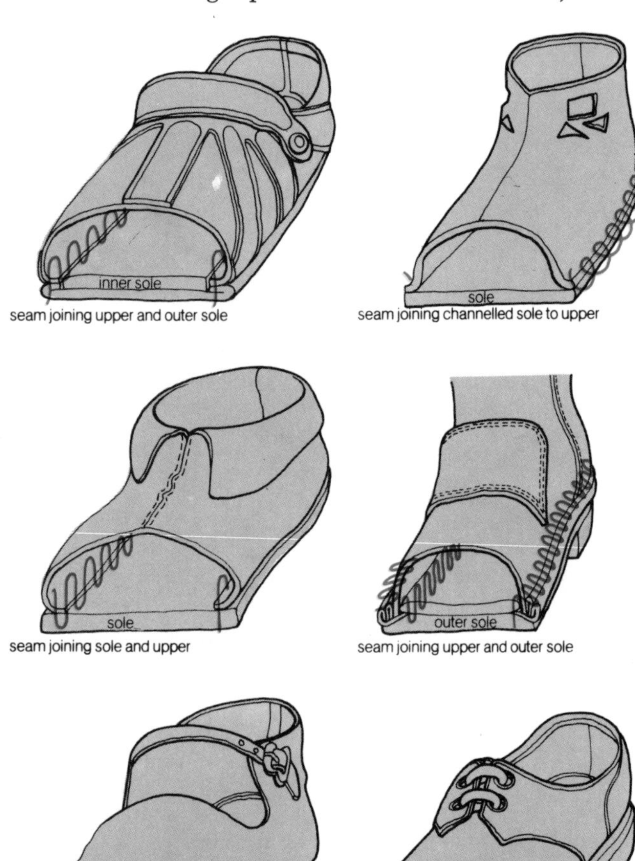

Above: methods of joining the uppers and soles of shoes from mediaeval times to the present day. The first four diagrams show simple mediaeval methods; the fifth a more elaborate boot from around the time of Edward III, incorporating a simple inner sole; and the last a modern welted shoe. This has both an inner and outer sole with a filling layer between the two, and would be found in high-quality footwear. Less expensive modern shoes use simpler methods, such as the use of adhesives, and many of the parts shown here are moulded together in one piece.

FOOTWEAR manufacture

Footwear originated when primitive peoples first protected their feet by wrapping them in the skins of animals they had killed for food. Making footwear achieved commercial status when some men began to make their living by making footwear for others. This had happened by about 2000 BC; there is a painting in the British Museum, from a wall in Thebes, which shows two men making sandals in about 1495 BC.

The patron saint of shoemakers is St Crispin, who was born about 300 AD. Shoemaking evolved as a craft most strongly in Europe, but it involved nothing like a factory until the nineteenth century; even then most of the work was done by 'outworkers' in their own homes. Mechanization began about 1860, with a strong lead coming from the USA; since 1950, it has spread rapidly, until today nearly all footwear manufacture is mechanized.

BASIC COMPONENTS A finished shoe or boot consists of the upper and the bottom. The upper consists of everything that can be seen above the bottom. Footwear manufacture, in very simple terms, consists of making the upper and then attaching it to

These three pictures show steps in the construction of high-quality made-to-measure shoes. Above left: a wooden last is made from a rough beechwood last to match exactly the foot of the customer. The last maker marks the last with a pencil where it is to be trimmed. Here he is shaving the marks off with a special blade called a last maker's knife. The pieces of leather are cut to match the last, using pieces of brown paper as patterns.

Below: sewing the pieces together in the closing operation. Then the upper is stretched over the last and fastened to it. The inner sole is attached, and has a piece of felt glued to it to prevent the finished shoe from squeaking.

Bottom: the shoemaker attaches the upper to the sole by hand stitching. The final stage is to put the heels on. This high-quality shoe would, of course, have leather soles and heels, for which stitching is the strongest form of assembly; but in lower-quality shoes the use of materials such as PVC, which can be ultrasonically welded together, has made this unnecessary.

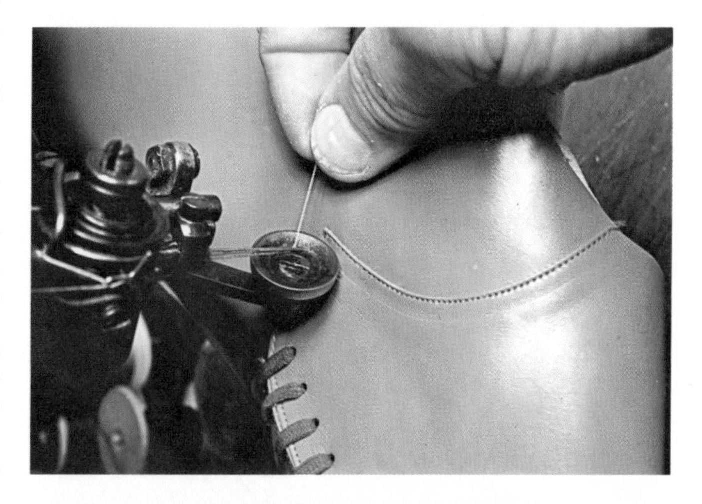

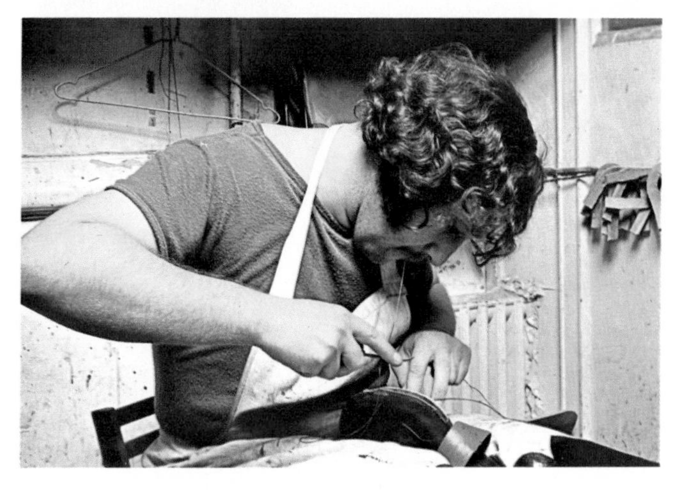

the bottom, in a variety of possible ways. The upper is still most often made from leather, especially in men's shoes, but a growing amount of manmade materials is being used, especially polyvinyl chloride (PVC) and polyurethane.

The various pieces that go to make up the upper are cut out in the clicking room. This operation used to be done manually by the clicker, who got his name from the sound his knife made as it slipped off the edge of the cutting board. Nowadays the clicker cuts out his pieces with a press. The pieces in an upper will vary according to the style and the design; they are normally sewn together on a machine in the closing room. An alternative method of closing, when the upper is made of PVC, is the use of high-frequency waves, which in effect welds the pieces together at the seams. Reinforcement of the toe and heel areas of the upper is provided by, respectively, the toe puff and the counter, or stiffener. These two components were traditionally made of fibreboard, but resin and resin impregnated fabric have become widely used. When the upper is closed, it moves on to the lasting operation.

LASTING The upper is stretched over a last in order to be processed through the rest of the shoemaking operation. The last is a foot-shaped form, formerly made of wood but nowadays often made of plastic. Another component is introduced at this point to keep the upper on the last: the insole is a bridge between the upper and the bottom. It may still be made of leather, but is more likely nowadays to be made of fibreboard or leatherboard. It is located on the upturned last temporarily by means of tacks; then the material around the periphery of the upper is stretched over the edge of the insole and fastened. Formerly this was done manually; the shoemaker would use pincers or forceps to pull the margin over and tacks to fasten it. Machinery has been developed to do this, first using tacks but more often today using adhesives. Attempts have been made since the early 1960s to build one machine to perform the complete lasting operation, but most factories use two or three machines to last the forepart, waist and heel sections.

Another component is associated with the insole: this is the shank, a girder between the waist area and

Above left: cutting the pieces for the uppers on a modern production line. The machine can cut several pieces at once.

Above: the closing operation is complicated here by the large number of pieces in the upper of this woman's shoe, needed to give the multicoloured effect.

Above right: the upper is stretched over the last and fastened. The next stage will be fixing on the sole. Although the shoe shown in this series of pictures is mass-produced, it is one of reasonably high quality, and so the process does not differ very much from that shown on p. 67, and a lot of hand work is involved.

Right: by contrast, here is a very highly automated process for the manufacture of sports shoes. Red PVC soles with a distinctive lightning flash design are injection moulded on to the canvas uppers of Dunlop Red Flash sports shoes; the same process adds the white locking strip around the edge of the shoe. Machines which weld or glue rubber or plastic parts together can achieve very high production outputs; here durability is not sacrificed, as the join between the PVC sole and the canvas upper is immensely strong.

the heel. It is obviously necessary in women's high heeled shoes, where there is a lot of curvature in the sole and heel. The shank is made of steel in such cases, but where the stresses are lower, as in men's shoes, it can be made of wood. Depending on the type of construction, the upper and insole may be subject to a number of ancillary operations before going on to the bottoming stage.

CONSTRUCTIONS Since shoemaking reached the factory stage, several types of construction have been developed. Today the constructions in use are welting, cementing, compression moulding and injection moulding. These terms define the method of attaching the upper to the bottom and sometimes the material from which the bottom is made. A welted shoe is most likely to have a leather sole attached by means of thread and tacks although it can be cemented. A cemented construction implies that the bottom is stuck on to the upper by means of adhesives; the bottom may be made of leather, resin rubber, crepe rubber, micro-cellular rubber, PVC, polyurethane or other polymer based material. Injection moulding implies PVC or polyurethane; compression moulded (or 'vulcanized') footwear implies rubber as the bottom material. An old construction, coming back to popularity recently, is that of string lasting; in this, the lasted margin of the upper is literally strung around its periphery and pulled tight over the insole.

BOTTOMING This operation can vary enor-

mously, according to the construction. Making a welted shoe implies a large number of operations; conversely, injection moulding or compression moulding implies very few. In moulding, the upper and insole assembly is presented to moulds which close over it, and the bottom material fills the mould cavity and is attached to the upper assembly. The machinery involved can be very large, in some cases involving as many as 48 stations which are rotated around injection points. Productivity from such units can be very high indeed, but such installations lend themselves to very high production runs and so are far more likely to be used where style changes are infrequent.

SPECIALIZED FOOTWEAR In addition to conventional footwear, there are other types with specialized applications. Industrial safety footwear contains steel toecaps capable of withstanding great loads; such footwear may also include a steel midsole to protect the feet of the workmen from sharp objects such as steel chips on the floor in a machine shop or nails on a construction site. Some footwear is made with antistatic properties, to be used by those working in plants where the smallest spark could set off an explosion or fire.

SHOEMAKING NATIONS No longer is shoemaking the province only of wealthy countries. With modern shoemaking machinery being exported all around the world, chiefly by Britain, Germany and Italy, it is simple for an 'emerging' nation to set up its own shoemaking industry. Those nations with the highest standards of living are the ones with the highest labour costs, so their home shoemaking industries are declining as nations such as Japan, Taiwan, China, Brazil and Hong Kong emerge as exporters of footwear.

FURNITURE manufacture

Furniture making is a craft as old as the earliest civilizations. Wood has always been the most popular material, being easy to work and aesthetically satisfying on account of its distinctive grains and colours. The Romans and the Greeks also used stone, marble and bronze to great effect; the range of materials which modern furniture makers have at their disposal is very wide, but the most important is still wood.

Furniture designing and manufacture was origally the work of individual craftsmen, but by the eighteenth century important furniture makers had teams of craftsmen helping them to carry out their ideas. From these small factories the making of furniture has developed into a highly organized series of processes. When a design has been approved to sell at a particular price, it is translated on the drawing board into details of construction methods; production flow charts are used to specify each process. In the factory, an abundance of conveyer systems are used, and several series of processes are used, converging appropriately and comprising an assembly line method.

WOOD PROCESSING Selected pieces of solid wood are kiln dried by the largest manufacturers to a moisture content which prevents the uneven strains of swelling and shrinking which would happen with untreated wood. Smaller firms buy the wood already dried. Modern furniture is made of pieces of solid wood in combination with particle board or chipboard (see MANUFACTURED BOARD). These are made of particles of wood held in a resin and pressed into standard sized sheets. The production line is usually divided into parallel lines, one for processing timber and one for the chip board.

In processing solid pieces of timber, the wood is first cut to approximately the right size, taking into consideration shaping or jointing which will be required later. Then the pieces pass through planers, which straighten one side of the board and plane the other face to the required thickness. Next it goes to a dimension saw bench, which cuts it into exact lengths and makes other cuts at required angles. The dimension saw is a table saw with a tilting angle adjustment. The final part of the preparatory stage for solid timber is the four or six cutter, which can trim all the sides of a board in one operation.

Another production line deals with the chip board. Panels of the required size are produced on a panel saw, which is normally set vertically. The saw moves, not the wood; the saw is mounted on a trolley which moves along the large sheet as required.

VENEERING Now that the two types of wood pieces have been prepared, they pass as required through the veneering shop. Veneer is a thin layer of expensive decorative wood which is glued on to a base of cheaper wood. Nowadays veneer is sometimes made of plastic with grains or colours manufactured

Below left: a Swiss craftsman at work in his shop. The highest quality furniture is still made by hand – a slow, painstaking process.

Below right: stapling upholstery. Tacks were originally used, but the stapler is faster.

Above: an assembly jig at the London College of Furniture, using hydraulic rams to hold the pieces together rigidly while the glue dries.

Right: selecting veneers. These must be carefully selected to match. Veneering is used to cover lower-priced base woods with thin layers of more expensive, exotic woods or other decorative materials.

into it; one of the most important types is melamine laminate, which is functionally ideal for kitchen or institutional furniture. Solid timber pieces are veneered because this makes possible an attractive finish at a saving in cost, since the base can be made of cheaper wood; particle board must be veneered if it is to be exposed in the final product, unless purchased by the manufacturer with an applied surface.

The veneers are cut to size by vertically set knife blades. There are also stitching machines for fastening pieces of veneer together where large pieces are required.

The adhesives used vary, but the most common is urea formaldehyde, a synthetic resin which requires a catalyst, usually ammonium chloride, to set or cure it. The curing is accelerated by heat treatment and it is common to use high frequency waves as the accelerator. The waves pass through the wood more readily than through the glue, and this resistance results in heat which speeds the cure. Veneer presses are used to ensure even adhesion; these are sometimes vacuum or dome presses. The vacuum press is a rubber sheet in a frame which is brought down on the work. An electric pump sucks out the air and the atmospheric pressure does the pressing. The dome press works the opposite way: a metal dome descends over the work and air under pressure is pumped into the dome. Whatever kind of press is used, heat treatment is used to speed the cure.

By using male and female forms in the presses, layers of veneer can be bent and formed into a laminated structure.

Edges of veneered panels can be veneered in the lipping press, a combination of vertical and horizontal pressures with high frequency curing.

FINAL PROCESSING Now that the two production lines have converged, the next series of processes are concerned with jointing, drilling, rebating and recessing. One of the most versatile of all machines used to make furniture is the double ended turnover. It includes circular saws and vertical and horizontal cutter heads. Once the machine is set up, it can work continuously. The pieces are carried forward by dogs (metal projections on an endless chain) and the cutter heads move in and out to produce tenons or double tenons, dovetails, trenches or mouldings, drilling, routing and other types of cuts in the appropriate places.

Other machines of importance are the mortiser, which cuts a mortice by means of a continuous chain cutter and a square hollow chisel and auger, and the router, which can be used for rounding corner or edge profiles and for producing decorative edge treatments such as spiral fluting. Spindle moulders, dovetailers, shapers and automatic lathes are also used.

SANDING After the shaping work is completed the various pieces are sanded smooth by belt or drum sanders. The sanders are moving tables on which the work is placed; the belt sander is a continuous belt pressed down on the work with pads. For complex mouldings the pads are profiled to match. The drum sanders are for panels and flat pieces; they are simply revolving drums covered with sandpaper. Soft pneumatic drums can also be used for certain shapes.

ASSEMBLY Furniture assembly is often divided into assembly and sub-assembly. The sides and top of a cabinet may be sub-assembled of frames and panels and then all the sub-assemblies come together for final assembly. (Such hollow box-like pieces of furniture are called carcass furniture.) Powered hand tools are used for stapling, nailing and screwing. Jigs and clamps are used to hold pieces in place during assembly. Jigs are patterns or forms using slots or raised barriers (positive stops) into which the pieces are inserted; clamps usually consist of rigid bars with sliding steel dogs. Where joints are glued, clamps and jigs are especially necessary and are used with high frequency curing. In some mass production situations, holes are drilled, glue squirted in and dowels inserted, all by machines.

FINISHING Either before assembly or after, the wood is stained if required and coated with varnish or paint. Mass production finishing is carried out with spray guns. Flat stock, including much knock down (KD) furniture which is assembled at home by the purchaser, can be finished on a curtain coater, a moving bed which takes the pieces through a series of spraying and drying operations. More expensive, richly finished pieces may have several coats of stain, sealer, varnish and so forth, applied by hand and hand rubbed or sanded lightly between coats.

UPHOLSTERY In mass production, many upholstery pieces can be cut simultaneously on a machine, but the upholstery itself is still a craft which requires skill and patience. A system of steel coil springs is tied into webbing, often made of burlap, and this assembly is tied to the frame. Sometimes the spring assembly is stuffed with foam rubber or some type of fibre stuffing. This assembly is then topped with padding made of foam rubber or cotton. Finally, inner covers and the outer covering of upholstery fabric are tacked to the frame of the piece. Economies which are often made today and which make possible some automation are the use of staples instead of tacks, steel bands instead of webbing, and clipping the springs directly to the wooden frame of the piece. When only a cushion is needed, these are often made of moulded sponge rubber or plastic foam. The outer covering can be made of leather or plastic as well as fabric.

OTHER TYPES OF FURNITURE Office furniture is often made of steel panels for durability; this lends itself to mass production. Lawn and garden furniture is often made of lighter alloys which can sometimes be worked by adapted woodworking machinery or can be readily assembled from cast components.

Some furniture, especially dining room sets, is made of aluminium or chrome plated steel tubing,

Far left: strips of veneer are stitched together to cover large areas.

Left: stain is wiped on and spread with cloths. This colours the wood, which will then be finished with varnish.

Above: stacking chairs being made in Denmark. Here, veneer is used for an eleven-ply construction, not decoration.

bent to shape and held together by screws. The tubing is bent in fixtures which are grooved to the diameter of the tube. Sometimes the hollow tube is filled with a resinous material to keep it from distorting when it is bent.

Plywood is also used in furniture manufacture. It is very strong, but large panels will not have rigidity; they can be given this by gluing two panels together with a solid wood panel or wood blocks in between. If a manufacturer requires a large number of plywood panels of a certain odd size or shape, they can be ordered that way direct from the plywood manufacturer. Plywood can also be made moulded into curved shapes for furniture; this is done by gluing the plies together in a steel form. This is then placed in a rubber bag, the air is pumped out to make a vacuum, and the assembly is subjected to heat and pressure.

Plastics have proved to be versatile materials for furniture today. Their applications include laminates, which are often bonded to plywood in an operation similar to veneering, plastic shells to take upholstery and all-plastic units. Such materials have produced radical changes in both furniture design and factory processes.

Both thermosetting (those which remain set after shaping) and thermoplastic (those which resoften on heating) types are used in furniture manufacture. Thermosetting plastics include melamine formaldehyde, urea formaldehyde, polyurethane, polyformaldehyde and the polyesters (which are the oldest plastics in furniture and are used reinforced with glass fibre). Thermoplastics include PVC, polystyrene, nylon, and acrylics such as Perspex (called Plexiglass in the USA).

Manufacturing methods include injection moulding, pressureless casting, extrusion, calendering and blow moulding. The raw material may be in the form of powders, granules or a viscous fluid.

GLASS

Glass is not a solid—it is a molten liquid of sand, usually with added limestone and sodium carbonate, which has been cooled to ordinary temperatures at which it becomes viscous and then stiff, with all the normal properties of a solid. This condition is called supercooling: when most liquids freeze they become crystalline, the crystal size depending on the rate of cooling. Glass, however, is unusual in its properties.

The liquid nature of glass can be seen by looking at old windows, which tend to sag downwards. Old Roman flasks which have been found resting under stones are sometimes squashed flat rather than broken.

Below: a German glass works in 1857. Sheet glass is being made by blowing a globe, drawing it out into a cylinder, cutting off the ends, slitting it up the side and flattening it out.

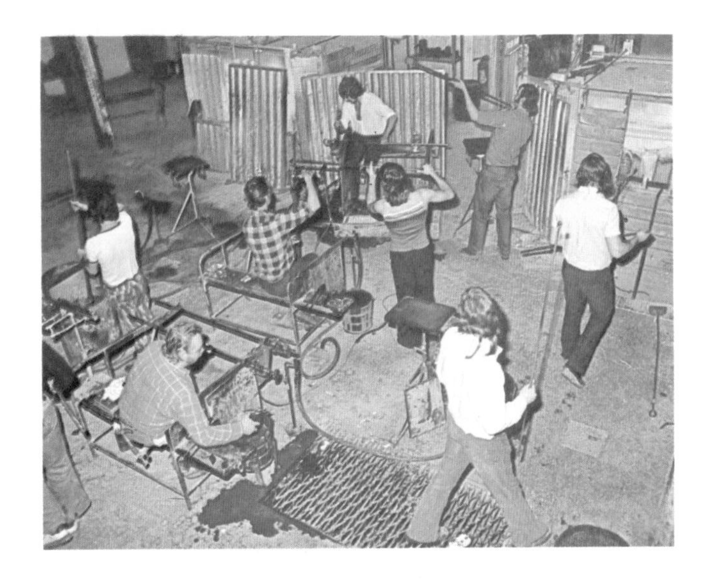

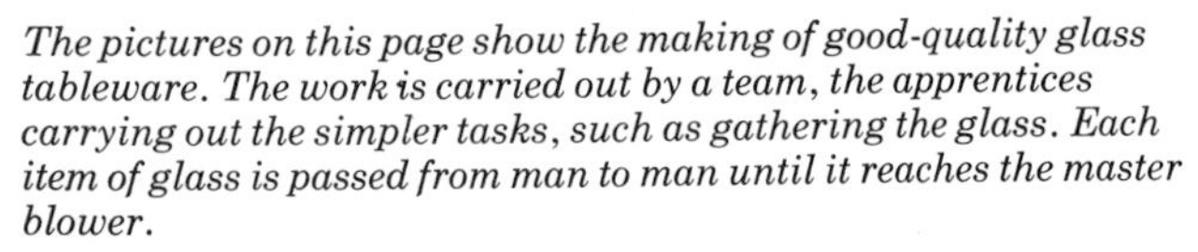

The pictures on this page show the making of good-quality glass tableware. The work is carried out by a team, the apprentices carrying out the simpler tasks, such as gathering the glass. Each item of glass is passed from man to man until it reaches the master blower.

Above: a general view: the furnace is at the top right; the master blower (seated) lower left. A gatherer takes molten glass from a pot in the furnace; the first blower blows a small bubble.

Above right: then he smooths it out on a metal slab called a marver.

Right: next, the shape of the glass is formed in a water cooled mould made of alder wood or graphite.

Below: now the master blower adds a stem, made of molten glass brought by a gatherer. After this the glass is passed slowly through a hot annealing lehr, gradually cooling over four hours.

Below right: finally, the waste glass is cut off by gas jets on a revolving table.

Glass can be made just from sand, which is crystalline silica or quartz, but the melting point is high (1700°C, 3092°F) and the result is still crystalline in nature. In this case, the material is known as a glass ceramic. Adding about 10% of lime (calcium carbonaate) and 15% of soda (sodium carbonate) produces a melting point of about 850°C (1560°F) and a much reduced tendency to devitrification (crystallizing). There are usually other components, their quantity depending on the type of glass needed.

HISTORY The first kind of glass made was faience, formed by melting the surface of sand grains together with soda or potash. This was used for making beads and small decorations at a very early date. In the second millennium BC the first true glass containers appeared. The Romans had cast window glass, but it was not particularly clear and was just used to let the light in but keep the weather out. The glass was cast in a flat sheet, and perhaps rolled while it was still hot to make it thinner. Although a few churches had glass windows as early as the seventh century, large sheets of transparent glass were not common until the seventeenth century.

Early processes involved either casting a sheet of glass, then rolling and polishing it, or blowing a globe of glass, then spinning it on the end of a rod so that it flattened out into a disc, already fire-polished and smooth. Old glazing using this crown or Normandy glass requires small window frames, some of which contain the characteristic central 'bullseye' mark from the rod. Alternatively, in the broad glass process, the globe was swung so that it extended into a cylinder some five feet (1.5 metres) long by 18 inches (45 cm) in diameter. The ends were then removed, the cylinder was slit lengthwise and flattened in a kiln.

This method, and a mechanized development of it, were used until the early twentieth century when two important processes were developed, the Fourcault (1904) and Pittsburgh (1926). In these, a ribbon of glass is drawn vertically from the glass furnace up an annealing tower by powered asbestos rollers which grip the ribbon as soon as it has cooled enough, a few feet above the surface. The tower, called a *lehr* or *leer*, allows the glass to cool slowly at a chosen rate. This is necessary to prevent stresses caused by the surface cooling too rapidly. The glass is transparent with hard fire-polished surfaces but exhibits some distortion.

THE FLOAT PROCESS Since its introduction by the British firm of Pilkington in 1959 the float process has become the world's principal method of flat glass manufacture. Previously, any flat plate glass had to be cast, rolled and polished to remove distortion.

The float process, unlike previous developments in flat glass manufacture, did not evolve from its pre-

The next four photographs show float glass making. In this, molten glass is floated out at a steady rate on to a bath of molten tin (below), spreading into a huge sheet which is cut up (right).

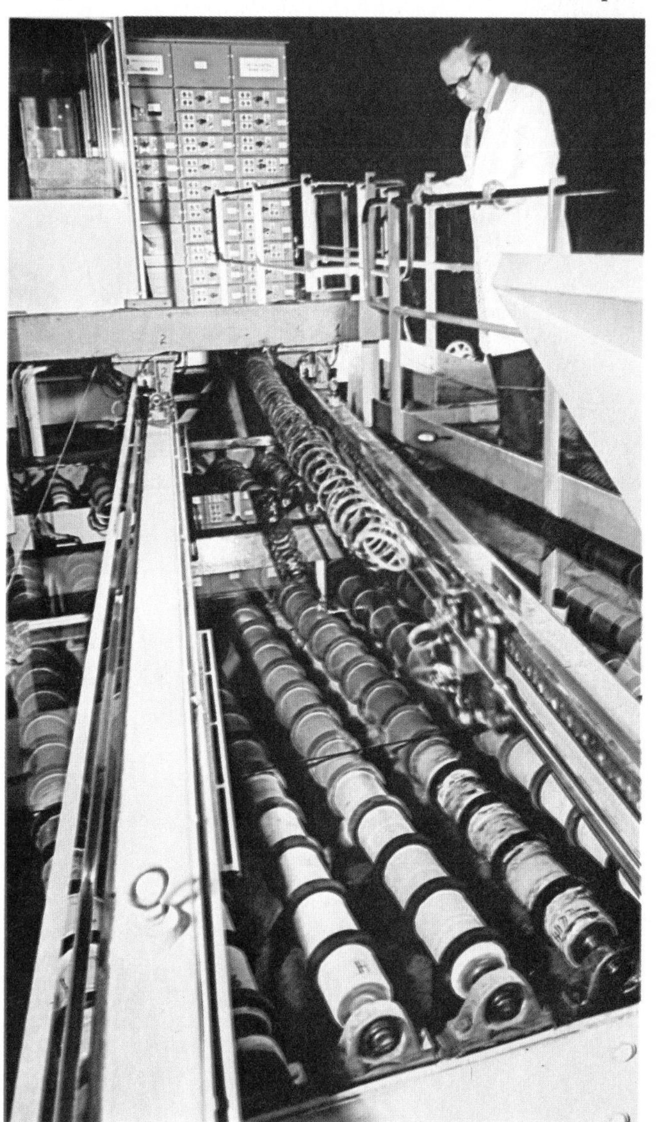

Above: the hot glass sheet, now no longer molten, coming off the end of the float bath and going to the annealing lehr.

Above right: after this, the glass passes under water jets which temporarily stress the surface to help the cutting process farther along the line. The scoring across of the sheet so that it can be snapped is shown on p. 75. The whole cutting process is automatic; it has to be, since glass emerges continuously and it is impossible to stop.

Below: a schematic diagram of the process, showing the ingredients being melted together, the float path and the long passage through the lehr on rollers. The float process depends on the fact that molten glass floats on molten tin but has a higher melting point, so that it can set while still afloat.

decessors. The advance was based on completely new technology.

To make polished plate, molten glass from a furnace was originally rolled into a continuous ribbon. But because there was glass-to-roller contact, the surfaces were marked. These had to be ground and polished to produce the parallel surfaces which bring optical perfection to the finished product. But grinding and polishing incurred glass wastage amounting to 20% and involved high capital and operating costs.

In the float process, a continuous ribbon of glass up to 11 feet (3.3 m) wide moves out of the melting furnace and floats along the surface of a bath of molten tin. The ribbon is held in a chemically controlled atmosphere at a high enough temperature for a long enough time for the irregularities to melt out and for the surfaces to become flat and parallel. Because the surface of the molten tin is dead flat, the glass also becomes flat.

The ribbon is then cooled down while still advancing across the molten tin until the surfaces are hard

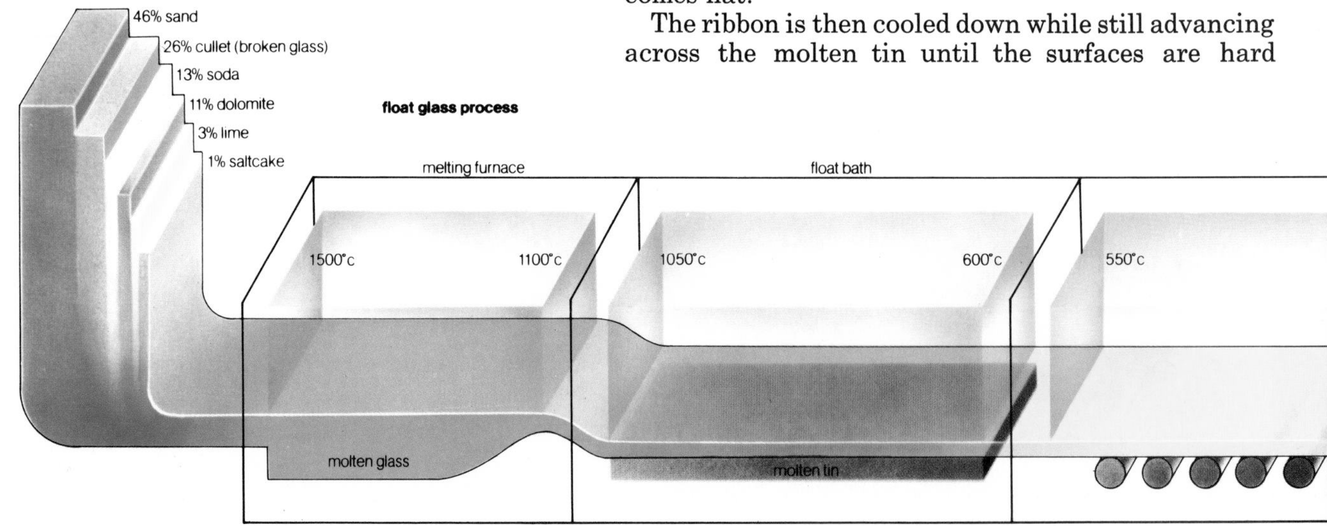

Left and bottom: photograph and diagram of an older sheet glass process by upward drawing from a pool of molten glass. Rollers pull up the glass; everything depends on surface tension. Now that float glass can be made in different thicknesses, this process has become obsolete.

enough for it to progress through the annealing lehr (in this case a tunnel) without the rollers marking the surface. The glass produced has uniform thickness and bright fire polished surfaces without the need for grinding and polishing.

After seven years work and 14 months of unsuccessful operations on a full scale plant, the float process was making glass about 6mm ($\frac{1}{4}$ inch) thick.

The natural forces within the float bath determined the glass thickness at about 6 mm—a fortunate phenomenon since 50% of the market for quality flat glass is for this thickness.

But the full potential of float could not have been realized without mastery of ribbon thickness. Just two years after float was announced, Pilkington could make a product half the thickness of the original float. The principle was to stretch the glass but in a gentle and controlled way so that none of the distortions arising in sheet glass processes could occur.

In the next three years thicker float was made. The spread of molten glass in the float bath is arrested and allowed to build up in thickness. The range of thicknesses now available commercially is 4 mm to 25 mm ($\frac{1}{6}$ to 1 inch) for the building trade, and glass down to 2.5 mm ($\frac{1}{10}$ inch) is manufactured for the motor trade.

OTHER TYPES OF GLASS Patterned glass is made by using rollers with surface patterns after the glass has emerged from the furnace. Wired glass, with a criss-cross pattern of wire set inside it, is made by rolling a ribbon of half the required thickness, overlaying it with wire and adding a further ribbon of glass which fuses with the bottom one. To make the glass transparent, the surfaces must be ground and polished. The advantage of wired glass is that it holds together when broken by impact or heat.

Optical glass is made in much the same way as other glasses with the important exceptions that the glass should be consistently homogeneous with no strain, striations or discoloration. Small differences in chemical composition or heat treatment can have considerable effects on the optical properties. Common glass usally contains iron oxides which discolour

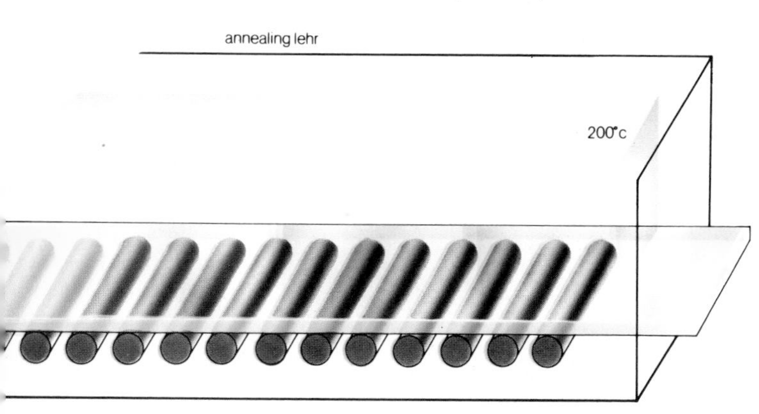

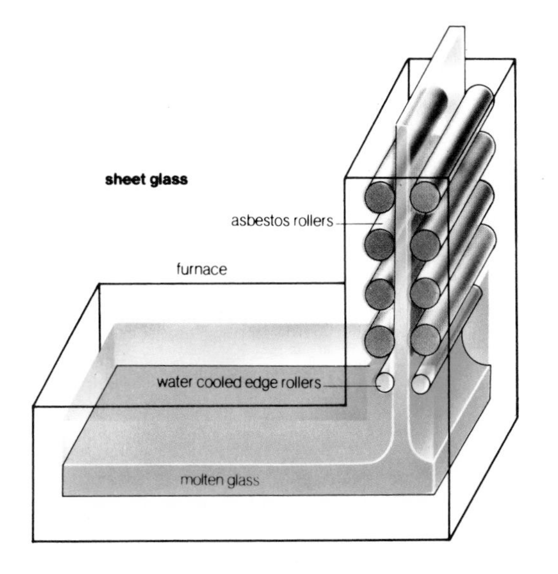

the glass. Sand used in the manufacture of optical glass is purer and the mix is suitably modified by the addition of various oxides of calcium, sodium, potassium, barium, or magnesium.

The batch is first melted, then the glass is refined by further raising the temperature. It is homogenized by being allowed to cool and then passed through a series of stirrers.

The optical properties of the glass are preserved by the critical melting, refining and stirring temperatures but the cooling rate is even more important, since this has an important role in determining the refractive index (light bending power) of the glass.

The cooling glass flows slowly down a delivery feeder and is sheared into globules or formed into sheet or slab for further processing. The globules are then moulded into lens blanks and passed into the annealing lehr.

Safety glass is made from annealed glass which undergoes a further process of toughening or lamination. Tempered glass has its temperature uniformly raised until it is just beginning to become plastic. The glass is quickly lifted out of the furnace, bent where necessary between matching tools and then cooled uniformly all over by jets of cold air blown forcibly on to it. The surface then becomes very much stronger than normal, withstanding shattering forces better and containing the glass if it does break.

Laminated safety glass is made by sandwiching a layer of clear or tinted polyvinyl butyral between two pieces of annealed float glass. The sandwich is gently heated under vacuum which evacuates all air from the laminate, and is then heated to bonding temperature under pressure. No adhesives are required. A number of layers may be used for increased strength.

If boiling water is poured into an ordinary table glass, the inner surface will heat and expand rapidly. Glass is a good insulator, so deeper layers will remain at room temperature. The resulting distortion may easily shatter the glass. In order to make glass ware which will withstand oven temperatures, a certain amount of boric oxide (B_2O_3) and alumina is added. This reduces the expansion coefficient of the glass by a factor of three. Pure fused silica is particularly resistant to thermal expansion, and so is used for the production of high quality optics.

CONTAINERS Glass has been used for making bottles and jars since its invention in the second millennium BC. Early vessels were built up on a clay or sand core, by dipping it in molten glass or by winding threads of hot glass around it. It was not until after 200 BC that the method of blowing glass vessels from a *gob* of molten glass on the end of a tube was introduced. This not only made better vessels, but was a much simpler and quicker process than the sand core method.

Early vessels were either free blown, resulting in a strong spherical shape, or blown into a mould, which gave the vessel an exact, repeatable shape and size and allowed designs to be impressed on it. Mould blowing forms the basis of the modern mechanical method of making glass containers, but for many centuries all glass was mouth blown, a craftsman's occupation which inevitably resulted in a slow rate of production.

It was only after 1882, when the glass industry in the north of England was paralyzed by a long strike, that mechanical methods came to be considered. The first machine was invented by a Yorkshireman named Ashley, and the design was greatly improved over the next 20 years. The fully automatic glass bottle machine was introduced by Michael Owens in the United States in 1903.

The normal composition of glass used for mass-produced containers is silica (70-74%), calcium oxide (9-13%), sodium oxide (13-16%) and alumina (0.5-2.5%). There is a trend towards increasing the calcium, and reducing the sodium content, which not only reduces the price of the glass (calcium oxide is made from limestone and is extremely cheap) but also increases its strength, making it more suitable for modern thinwalled containers.

Glass naturally has a slight greenish tint from the

Left: the difference between laminated and toughened windscreens. When hit by a stone, an laminated screen cracks but is held together by its plastic inner layer, while a toughened screen crazes and becomes almost opaque.

Left: the mass production of glass bottles. They are being transferred from an automatic moulding machine on to a conveyer which takes them to the annealing lehr just visible in the background. Although they look as if they are full of whisky, this is simply because they are red hot.

presence of iron oxide in the raw materials, so clear glass is decolorized by adding small amounts of selenium and cobalt oxide. Coloured glass is frequently used, sometimes for decoration but often in containers for liquids such as wine or beer, which are affected by light. Green glass is made by adding varying amounts of iron, manganese, chromium and nickel oxides, depending on the shade required. An amber colour comes from a mixture of carbon, sulphur and iron. Blue is obtained by adding cobalt oxide or copper oxide; red from selenium and cadmium sulphide. Opal (milky) glass, which can be white or coloured, is made by the addition of fluorides or phosphates and alumina.

Glass furnaces are generally run for several months or years continuously, so most manufacturers cannot supply the entire range of colours. Glass containers are manufactured around the clock to suit the continuous operation of the furnace. The furnace itself may hold from four to 100 tons of glass, and operates at temperatures up to 1590°C (2890°F).

The raw materials are weighed and mechanically mixed so that the minor constituents are evenly blended. They are then taken by a conveyor to be charged into the melting chamber of the furnace, where they fuse and mix into a homogeneous liquid mass.

From the melting chamber the glass, which flows as freely as water at this temperature, runs to the second zone of the furnace, the working chamber. This is a reservoir where finished glass is allowed to to cool slightly, becoming slightly viscous and suitable for forming.

The furnace is generally mounted on a raised platform above the forming machines. Glass flows down through a long feeder channel in which it is brought to exactly the right temperature for forming into gobs. The weight of a gob, which must be constant to produce identical containers, depends on the viscosity of the glass and hence on its temperature.

A machine picks up the gobs and drops them into a passing line of rough moulds, where by pressing or compressed air blowing they are made into an intermediate shape called a parison, roughly the same shape as the intended container but smaller and thicker.

Another system uses suction to form parisons by drawing up molten glass from revolving pots and severing it with a blade.

Each parison is then transferred to a finishing mould and blown to its final shape. Moulds are generally made of cast iron, and are in several pieces so that they can be opened to let the container out. A plain jar mould has left and right halves and a separate bottom. Apart from the manufacturer's standard moulds, special moulds of unusual shapes are often made to a customer's specification: for example jars for cosmetics.

The containers soon cool enough to stand on their own, and are released from their moulds and conveyed to the lehr. Finally, they are inspected and faulty containers rejected; these are used as cullet—added to the raw materials when making a fresh batch of glass.

This process can make well over 200 containers a minute; other, semi-automatic processes are in use at smaller factories, using small pot furnaces holding less than a ton of glass, from which the glass is gathered by hand on an iron rod. There may be up to twleve pots per furnace, and each can hold a different colour if required.

Some high-quality bottles and glasses are still mouth blown.

GUITAR construction

The guitar is a member of the fretted instrument family in which the pitch of the strings is altered by pressing them down behind 'frets', which are metal strips attached to the fingerboard. Other members of the same family include the banjo, mandoline, bazouki, balalaika and ukulele. Guitars of whatever type—classic, flamenco, plectrum, acoustic, 12-string, Hawaiian or electric—are all descendants of the instrument which evolved over the centuries, mainly in Spain.

HISTORY Among the earliest evidence of the guitar is the illustration of the 'guitarra latina' in the *Cantigas of Alfonso the Wise* (Spain, about 1270 AD). This is a true guitar; that is, it has a body with incurved sides and a longish neck fitted with frets, carrying 4 strings attached to a bridge fixed to the lower part of the soundboard, which is pierced by a single central soundhole.

Where the ancestor of this guitar came from is still uncertain, due to lack of reliable research, but a

Hittite carving (about 1350 BC) depicts an instrument with strong guitar characteristics. Later Persian ceramics show another guitar-like instrument, the Târ, and it is possible it spread as far as the Pillars of Hercules (Straits of Gibraltar) in ancient times and became established in the Iberian peninsula. Even the name of the 4-string Târ, 'chartâr', could have become 'guitarra' in Spanish, particularly since the first guitar of which there is evidence was a 4-string guitar.

In his *Declaración de Instrumentos* (1555 AD) Fray Bermudo states that the guitar had 4 strings tuned G D g b (capitals are used for the bass strings, and small letters for the treble strings) for playing 'popular music' or A D g b (the middle 4 strings of the modern guitar) for 'serious and composed music'. By 1586, when Juan Carlos y Amat wrote his method for the *Guitarra Espagnola* it had acquired another string, the first (top) e string of today's guitar. At the end of the 18th century the sixth (bass E) had become a permanent addition and E A D g b e was established as the standard tuning for the guitar.

Until this time the guitar had been strung with pairs of strings, a custom that has survived on some versions of the guitar in Latin America (and on the mandoline). In fact the popular modern 12-string guitar is almost certainly derived from a large double-strung guitar popular in Mexico. However, single strings were generally adopted around the end of the 18th century, the top three being of gut and the lower three of silk floss covered with a winding of silver or copper wire. Nowadays the three treble strings are of extruded nylon and the basses, of wire-covered nylon floss.

The early years of the 19th century saw a great upsurge of popularity for the guitar throughout Europe and virtuoso guitarist-composers such as Fernando Sór, Dionisio Aguado (both from Spain), Carruli, Carcassi, Giuliani, Molino (from Italy) and Napoleon Coste (Corsica), became the objects of critical acclaim and popular esteem. This fashion for the guitar created a great demand for instruments which was met by fine luthiers (builders) such as Lacote and Grobert (Paris), Stauffer (Vienna) and Panormo (London). Run-of-the-mill guitars could be bought for as little as 7s 6d (then $1.50) but a Panormo guitar 'in the Spanish Style' cost from '2 to 15 gns' (then $8.40 to $63.00).

In the 1860s and 1870s, however, the guitar began to wane in popularity as the big sound of the piano-forte and the large orchestra caught the public fancy. In 1833 a German guitar maker, Christian Friedrich Martin, emigrated to the USA and set up a workshop in New York, later moving to Nazareth, Pennsylvania, where C F Martin and Co. continue to make fine guitars.

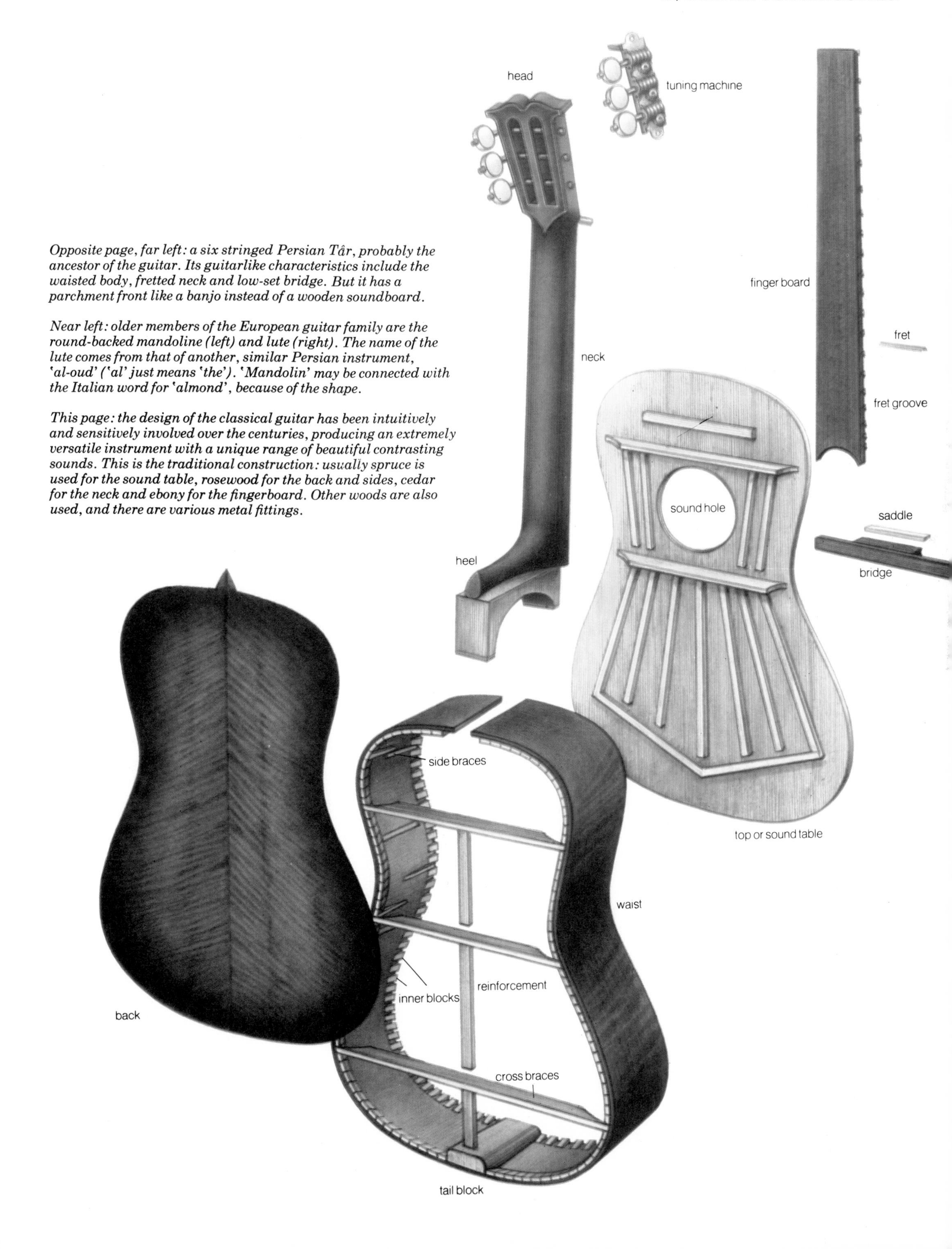

Opposite page, far left: a six stringed Persian Târ, probably the ancestor of the guitar. Its guitarlike characteristics include the waisted body, fretted neck and low-set bridge. But it has a parchment front like a banjo instead of a wooden soundboard.

Near left: older members of the European guitar family are the round-backed mandoline (left) and lute (right). The name of the lute comes from that of another, similar Persian instrument, 'al-oud' ('al' just means 'the'). 'Mandolin' may be connected with the Italian word for 'almond', because of the shape.

This page: the design of the classical guitar has been intuitively and sensitively involved over the centuries, producing an extremely versatile instrument with a unique range of beautiful contrasting sounds. This is the traditional construction: usually spruce is used for the sound table, rosewood for the back and sides, cedar for the neck and ebony for the fingerboard. Other woods are also used, and there are various metal fittings.

In 1921 the Martin Co. produced larger, sturdier instruments to be strung with steel strings and played with a plectrum. These instruments, like those made by other North American makers, were still based on the old Spanish guitar, but a few years later the Gibson Company started to produce large steel-strung guitars with arched fronts and backs carved from solid wood, which became known as 'cello guitars and which produced a crisper, more penetrating sound suitable for use in jazz and in dance-bands. During the 1930s magnetic pickups were fitted beneath the strings of cello guitars thus producing the electric guitar and paving the way for the modern solid and semi-solid instruments. The musician who first saw the new potential for the electric guitar in jazz was Charlie Christian, a black musician from Oklahoma, who died tragically young of tuberculosis.

THE CLASSICAL GUITAR In the meantime things had been happening to the guitar in Spain where in 1850 a guitar maker, Antonio Torres, egged on by the guitarist Julian Arcas, evolved a fuller bodied, more powerful yet still refined guitar which became the prototype for all modern classic guitars. An additional

advance came from the guitarist Francisco Tarrega who advised a new rational technique which allowed the player to exploit the improved qualities of the Torres model guitar and which became the basis of modern guitar technique.

In the year that Tarrega died (1909) the sixteen year old Andrés Segovia gave his first public performance. Through his artistry, the strength of his personality and a burning desire to advance the guitar, he was largely responsible for the new renaissance of the classic guitar and its full acceptance as a serious instrument in the world of music.

ELECTRIC AND ACOUSTIC GUITARS Parallel to the post World War II revival of the Spanish and classic guitar, has been the growing popularity of other forms of the guitar such as the electric guitar in the jazz, rock and pop scenes and the flat-top acoustic guitar which has become the folk and blues instrument *par excellence*, whether single or double-strung.

Basically there are only two types of guitar, acoustic and electric. The sound of the electric guitar is produced by the string vibrating above coil and magnet. The string must be steel or some other suitable magnetic material and under each is positioned a small permanent magnet surrounded by a coil of wire (the pickup). As the string moves (vibrates) the surrounding magnetic field is distorted, which induces a voltage in the coil. The resulting electrical signals are fed into an electronic amplifier where they are converted into sound. The body of the electric guitar plays little or no part in the quality or quantity of the sound. Volume and tone controls are used to regulate the power and quality of the sound which comes from a speaker (or speakers) linked to the amplifier.

The acoustic guitar, classic, folk-blues or 'cello built all rely on the vibrations of the strings passing

Far left: three types of guitar. From left to right, a Spanish classical guitar, a 'cello' guitar with arched front and back and f-holes instead of a central sound hole, and a Dobro with metal inserts which give it a resonant, metallic sound favoured by many blues singers.

Below: the use of holography makes the pattern of vibrations in a soundboard visible as light and dark bands. This guitar was sounding a note at a pitch of 254Hz, just below a standard middle C, which is 261Hz.

via the bridge into the soundboard which is then set in motion. The sound produced is amplified by the sound-box. The shape, size, and construction of the body (soundboard and back and sides) determine the quantity and quality of the sound.

CONSTRUCTION Antonio Torres was followed by a group of Spanish guitar makers who established firmly the principles which were gradually adopted by non-Spanish luthiers, most noted of whom was the German maker, the late Hermann Hauser. Nowadays there are fine guitar makers in various countries throughout the world. Every maker has his own version of the Torres design but they all remain close enough to the original to see from whom the tradition stems.

The basic idea behind this design is that the system of fanstrutting beneath the soundboard allows it to vibrate as fully as possible while still maintaining and supporting it. Torres was not a scientist or an engineer. He was a fine craftsman who understood wood and was devoted to making excellent guitars and it must be admitted that the best guitars have been made by craftsmen who have sensed intuitively what needed to be done with each piece of wood.

Attempts have been made to assess scientifically the resonant qualities of woods with a view to improving the guitar but the best instruments are still made of the woods which tradition has proved to be best, namely spruce for the soundboard, rosewood for the back and sides, ebony for the fingerboard and Cuban cedar for the neck. The bridge is usually of rosewood. Other woods such as maple, sycamore, algarrobo and walnut are suitable for back and sides but rosewood remains the first choice.

The Flamenco guitar has cypress for the back and sides, a light brittle wood which contributes to the bright tone required by a Flamenco player. Ideally wood for guitar construction, especially the spruce for the soundboard, should be quarter cut, that is, cut from the trunk like slices of a cake, so that the annual rings pass through the wood at right angles. Flat-top acoustic, steel-strung guitars have sound-

Other stringed instruments are constructed in different ways. On the left, a rebec, a mediaeval ancestor of the violin, was made like a dugout canoe, while a real violin is made from delicately arched, wafer-thin pieces, as shown below.

boards of spruce and backs and sides of rosewood or mahogany. The strutting of the soundboard is simpler and sturdier than on a classic guitar. The best 'cello guitars have soundboards carved from spruce and backs and sides of maple. Instead of a soundhole they usually have f-holes like a violin and two longitudinal struts, one on the bass and one on the treble side of the soundboard.

The Hawaiian guitar differs from other guitars. It is played resting flat across the knees, soundboard uppermost, and the pitch of the strings is changed by sliding a round or flat piece of metal, 'the steel', up or down the fingerboard producing, incidentally, the wailing sound typical of the Hawaiian guitar. The right hand fingers are equipped with metal or plastic finger-picks. Various tunings are used which allow chords to be played by simply placing the steel at right angles or diagonally across the strings. A sophisticated version of the Hawaiian guitar is the 'pedal guitar', an important component of the Country and Western group on which various tunings can be selected by the use of a pedal.

In recent years, rock, pop, and folk guitarists have got into the habit of employing different tunings which facilitate special harmonic sequences or personal, or traditional styles. It has always been customary to retune the sixth string to D and sometimes the fifth to G on the classic guitar, but generally speaking the traditional tuning remains the most useful and characteristic.

With the exception of Spain and Latin America, where the guitar has always been the popular folk instrument, it has tended to be an instrument of fashion which has resulted in its development being spasmodic. Today it is enjoying an unprecedented boom and it remains to be seen whether it has achieved a permanent position in the world of music in general.

INK manufacture

The oldest written document in existence is an Egyptian payrus dating from 2500 BC. The development of inks since then has led to an important technology in modern printing inks; large manufacturers conduct their own chemical research to keep up with the demands of the printing industry. The technology of printing continues to develop; the scope of it can be seen in the fact that inks in each colour of the basic spectrum are available in as many as a hundred shades.

Early inks were made by dispersing carbon in the form of lamp black or soot into water or oils. From the seventh or eighth century, metal gall inks began to appear; these are a mixture of galls and metal salts. (Galls are growths on trees caused by attacks of various kinds of insects on the plant tissue.) The galls reacted with the salts and gave a black or brown solution. Later the galls were replaced with tannin, and indigo, a natural dye, was added. Indigo was in turn replaced by synthetic coal tar dyes.

WRITING INKS Writing inks are coloured with soluble dyes and dry by penetration of the paper and evaporation of solvent. They are manufactured by

Below: paper chromatography shows the ingredients of black ink spreading at different rates.

Above: a device for testing the colour density of ballpoint inks. The ink must be tested in an actual ballpoint, since only a small amount gets past the ball housing.

stirring the dyes into the solvent, any insoluble material being finally removed by filtration or by passing the ink through a centrifuge.

In modern blue-black fountain pen ink, the writing trace is coloured blue by the dye. On exposure to oxygen in the air the blue dye fades and the metal tannate oxidizes to a black colour. This ink is permanent and will not fade from the paper; washable inks are simple solutions of colouring material in water containing a little preservative.

The fibre tip of a fibre tip pen consists of a bundle of textile fibres bound together by a resin; enough air space is left between the fibres for the ink to flow through them by capillary action. The ink is similar to ordinary writing ink but has a high humectant (moisturizer) content to prevent the tip from becoming clogged with dry ink.

In the ballpoint pen, ink is transferred to the paper from the surface of a ball of 28 to 40 thousandths of an inch (0.7 to 1.0 mm) in diameter. Because of the small size of the ball and the space between the ball and its housing, the volume of ink delivered to the paper is small, so the ink contains a large amount of colouring matter to make a legible trace. For comparison, the ink contained in a small ballpoint pen refill will draw a continuous line 1500 to 2000 yards long; the same volume of fountain pen ink would draw a line only 80 to 100 yards long. To prevent the ink from drying on the ball, and solid matter from jamming between the ball and its housing, the ink contains very slow drying solvents and is carefully filtered. Since component dimensions vary from one pen design to another, ballpoint inks have to be designed to match the pen in which they will be used.

PRINTING INKS Printing inks are designed according to several variables, such as the material to be printed, the printing process to be used, the drying method, the type of press, required permanence and colourfastness and so forth. The materials printed upon range from cellophane to steel, including a great variety of types of paper. The three main processes of printing are letterpress, which uses raised surfaces to carry the ink to the paper; gravure, which uses recessed surfaces (the ink must be pushed into the recesses); and lithography, which uses a flat surface, parts of which are greased or otherwise treated so that the ink does not stick to them. Letterpress processes can be further broken down into news, magazines, book and jobbing printing, in roughly descending order according to the speed of the presses.

Printing inks will gel in storage, and must be able to be liquefied by mechanical agitation. They must not gel while on the press, which might be a period of several hours. All the printing processes involve pressure of the machinery against whatever is being printed, so if the ink is too fluid it will squash and make a messy printing job. This is prevented by ensuring that the viscosity is correct. The coating of ink is thin, which means it must be heavily pigmented. These requirements mean that in general printing inks are much heavier in consistency than writing inks, but there is a wide variation among them.

The pigments used are much the same, but the liquid part of the ink (the vehicle) varies according to the job. Various resins, drying oils and often solvents are used. For example, jobbing inks, used for printing stationery, tickets and the like, are thick and slow drying; the vehicle will be a mixture of mineral oil, resin and litho varnish (specially treated linseed oil). On the other hand, news printing can use just mineral oil because it is applied to more absorbent paper on much faster presses.

Inks used in gravure processes must not be too tacky, because the machinery wipes the plates clean and the wiping action must not pull the ink out of the recesses; at the same time, the ink must have enough body to fill the recesses completely. It must not react with traces of acids used in etching the plates, while the finely ground pigments should be 'soft' so that the copper cylinder is not scratched. Inks used in lithographic processes must not dissolve in water or they will 'bleed' on to the non-printed area.

Rough paper requires more ink than smooth paper. A given volume of coloured ink will not cover as much as black ink, the coverage varying with the colour. There are so many shades of coloured ink that no satisfactory method of classifying them has been devised; matching must be done visually or by using a spectroscope.

Most printing inks dry partly by penetration and partly by reaction of an oil or synthetic resin with oxygen in the air, but there are many variations. Heat-set inks are often used in high-speed presses; they solidify when heated and are made with solvents which are volatile at high temperatures instead of drying oils. Some of these contain polyvinyl chloride (PVC). Cold-set inks are applied at a relatively high temperature and dry when they cool. Steam-set inks dry by precipitation and instead of drying oils the vehicle comprises a glycol solvent and a resin which is insoluble in water, so that on absorbing water the pigment and resin precipitate and set on the paper simply by keeping the press room humid.

Metallic inks have particles of metal suspended in them; they are used for printing on metals and packaging. Sometimes inks are made deliberately magnetic, for printing banknotes, cheques and business forms which are sorted or 'read' by machinery. Fluorescent inks are designed for special colour brightness, and can be coated to prolong the period of brightness. Some inks are designed to be applied through a a screen, and some are applied electrostatically, that is, pulled through a mask by means of an electric charge.

As can be seen from all this, the variety of printing inks available is very wide and growing daily. Printing inks are coloured by insoluble pigments, and after mixing must be passed through a roller mill to ensure that the pigments are dispersed thoroughly throughout the vehicle. This machine consists of rollers revolving closely together at different speeds. The particles of pigment are crushed and sheared until they are coated with the oil or other vehicle, and small enough so that they will not cause wear of the printing surfaces. The mixture is passed through the rollers several times, and additives can be added during this process or afterwards.

LEATHER making

The making of leather from animal skins is one of the oldest accomplishments of mankind, and its origins are lost in history.

The hide of an animal contains bundles of tough protein fibres, and when the animal is alive these fibres contain moisture and bacteria To preserve the hide, the bacteria must be killed by salting or drying. The hydrophilic (water absorbing) fibres, called collagen, must then be chemically treated to prevent them from putrefying. This is called tanning. Prehistoric men wrapped themselves in animal skins

Below: in Kenya, cattle are important in the economy, and leather is an important industry. This is 'flensing', the first step in making leather: the scraping of loose skin from the underside of the hide. Next, it will be soaked in a vat for a couple of weeks.

Above: this picture was taken in Ethiopia, but much the same methods are used elsewhere. It shows leather being stretched on a frame to flatten it.

Below: 'staking' – pulling the leather across a sharp edge—is an ancient technique. It keeps the bundles of protein fibres in the hide flexible while hastening curing.

for magical reasons, hoping to make themselves better hunters; they must have discovered that by rubbing fat from the body of the animal into the hide it would last longer and be more comfortable to wear. Trying to dye skins by applying various pigments to them, they presumably discovered that certain substances had a preserving effect. Before the beginning of recorded history, hides had become a valuable trading commodity.

A great many ways of preserving hides have been developed over the centuries. Hides have been smoked, salted, treated with urine, soaked in pits with a solution of animal dung, beaten and scraped across pointed sticks. Talc, flour and many other substances have been beaten into skins to replace the natural moisture. Prehistoric men (and Eskimos in more recent times) have chewed skins in order to remove hair and bits of flesh and to make them soft. The chewing action, by removing fat upon which bacteria feed, may help to preserve hides. Treating hides was always hard work until machines began to take over in the nineteenth century, because of the necessity of beating the treatment into the skin. Tanners, despite the value of their work, were often forced to live outside the city limits because of the unpleasant smells and diseases associated with some of the processes.

A process which allows the hide to be used but does not make it water resistant is called pseudo-tanning. A true tanning process chemically penetrates the atomic structure of the fibres, preserving their toughness and durability and making them water resistant. Tanning may have been discovered when a hide was left in a puddle containing leaves, bark, acorns and other vegetable matter containing tannic acids. The bark of hemlock and mimosa; the wood of the chestnut tree and the *quebracho* (a South American tree); valonia, the cup of an acorn which comes from Europe and Asia minor; myrabalams, the unripe prune-like fruit of an Indian tree; and many other vegetable parts all contain tannic acids.

For thousands of years men have also practiced mineral tanning, using alum salts. In the nineteenth century chrome tanning was invented, using the salts of the metal chromium; this process took only a few days instead of weeks, and resulted in a hard, stiff, blue-coloured leather which was comparatively waterproof. Ways were found to make the chrome-tanned leather soft and pliable by treating it with soaps and oils, and today the uppers of leather shoes are made of fully chrome-tanned leather or semi-chrome (partly vegetable and partly chrome tanned). Nowadays the full range of technical and chemical knowledge is applied to produce an extremely wide range of leathers with great efficiency; one of the recent developments is the use of aluminium salts, which produces a white leather that is lightfast; another is the use of solvents in the drying process, which is very fast.

FACTORY PROCESSING Hides are often refrigerated, salted or packed in barrels of brine to cure them

Left: this machine reduces the thickness of the hide to the required uniformity by 'shaving' it.

Below left: after tanning the hide is pasted to a glass plate for drying, a method known as paste drying. It travels through a heated tunnel.

at the meat-packing plants as soon as they are removed from the animals. Some hides are partially dried to cure them, but not top quality stock: drying lowers quality and causes problems in later stages of processing. When the hides arrive at the tannery, they are trimmed to remove useless edge pieces and so on; then they are sorted according to size, weight and thickness.

Hides are next soaked in water in revolving drums, to which bactericides and detergents may be added. This softens them and prepares them for further treatment. It also removes some of the food proteins which could support the growth of bacteria; the soaking must be carefully monitored to prevent the production of these bacteria, which would ruin the stock.

If the hair is of value, the hides are dehaired by means of a chemical spray or a paste of sodium sulphide applied to the underside. This penetrates to the roots of the hair and loosens it. If the hair is not valuable, the method used is liming (soaking in an alkaline lime solution). This is done in rotating wooden tubs or in stirred vats called paddles. It loosens the hair, destroys the epidermis and also loosens the fibre bundles. Sodium sulphide is added to assist removal of the hair and epidermis; a scudding machine then scrapes off the loose hair with dull blades.

The next treatments depend on the type of leather and the processes to follow. The hides are delimed by washing and then soaking in an acid solution, followed by bating, a mild process which treats the hides with enzymes to remove some of the collagen, softening them. (Bating may be omitted for firm leathers such as those used for shoe soles.) The hides are then pickled in a solution of salt and sulphuric acid, especially those to be chrome tanned. In modern tanneries, these processes are carried out in a single unit process.

TANNING There are three main types of tanning in use today: vegetable tanning, mineral tanning (mostly chrome tanning) and oil tanning, for leathers such as chamois.

Vegetable tanning takes the longest. The hides are soaked in a tannic acid solution for as long as several weeks, being moved to progressively stronger vats of solution. Heavy leather, such as that intended for shoe soles or industrial belting, gets the longest treatment; light leathers such as calfskin, used for bookbinding and luxury goods, are thoroughly delimed and sometimes pickled before tanning. Light leathers are tanned in as little as twelve hours.

Mineral tanning takes much less time and gives an extremely durable product, although it results in colouring of the stock, which means extra treatments to conceal the colour. There are single bath and double bath methods; sometimes the hides are tumbled in drums with paddles to keep them agitated. Nowadays the temperature and strength of the various solutions is carefully monitored by instruments.

Oil tanning is a direct descendant of ancient methods, using fish oil to replace the natural moisture. The leather is dried beforehand, partially and carefully, then beaten in a machine while sprinkled with the appropriate oil. The process is repeated until all the moisture has been replaced, and then the oil is decomposed by means of heat.

After tanning the hides are washed and wrung out; then they are split or sliced on a band knife to give uniform thickness. Retannage may then be applied to produce required characteristics, such as suppleness for gloving or upholstery leather. If necessary, dyeing and fat liquoring follow. Dyeing is necessary at this stage if transparent finishes are to be applied; for example, aniline dyes for upholstery. In fat liquoring the hides are treated with emulsions of natural oils or greases to reduce the degree of coalescence of the fibre bundles in subsequent drying.

DRYING Heavy leathers are dried slowly, although chrome leathers are so tough that they can be dried in a few hours. Sometimes the leather is hung on hooks in a stream of warm air; the grain side is coated with oil to force the moisture to leave through the underside. Most leather nowadays is dried in moving 'tunnel' type conveyers. One way to prevent shrinkage is to stretch the leather on frames by means of clips or toggles before passing it through a drying cabinet.

Now the leather is stiff, so it is brushed or sprayed with a solution of water and soap and allowed to stand to *condition*, or achieve the desired moisture content. Some stock intended for industrial purposes is subjected to stuffing (additional grease impregnation) to render it water repellent.

Many leathers are then subjected to a staking action, the equivalent of the ancient method of pulling it across the ends of pointed sticks. The modern method is flexing and manipulation in machines to loosen the fibres. Final drying follows, often using a vacuum drying cabinet.

FINISHING Leather is buffed to bring it to uniform thickness by revolving steel cylinders covered with abrasive paper. For suedes, a fine nap is produced by high speed emery wheels on the flesh side of the leather. Leather for shoe soles is pressed by heavy machinery to give it the required firmness. Patterns are imposed on the grain side by means of heated engraved plates under pressure. Box or willow grains are made by rolling the leather over on itself between cylinders revolving in the same direction. Glazes are applied by means of heated glass or steel cylinders. Leathers are dyed using pads or brushes. Resin fillers and pigmented or clear lacquers are also applied, usually by spraying. When the finishing operations are complete, the leather is shipped to factories and manufactured into consumer goods.

SYNTHETIC LEATHER Among the many applications of leather, footwear is the most demanding in terms of the strength, durability and flexing characteristics required. Footwear leather also requires the ability to absorb and dissipate moisture during wear in order to maintain foot comfort and hygiene. In order to meet the increasing demand for leather, which cannot be met from natural sources, many types of synthetic leathers are now produced to meet the particular requirements of a variety of applications which, apart from footwear, include upholstery, garments and luggage.

Cheap leather substitutes first appeared in the 1920s with the introduction of 'oilcloths' and 'leathercloths', which were augmented by vinyl coated fabrics in the 1940s. These early materials cracked too readily when flexed and had a relatively short life, and none of these early materials were permeable to air or water vapour. Since the 1950s many alternatives have been introduced which have proved more satisfactory.

Synthetic leathers can be grouped into those materials now known as poromerics, which attempt to reproduce all the properties of natural leather including 'breathability', and those which simply attempt to look and feel like natural leather, without any attempt to build in foot hygiene promoting properties. The latter group are mainly plastic coated fabrics.

POROMERICS The first poromeric to be introduced was Corfam, produced in the early 1960s by the DuPont Corporation in the USA. Although it did not prove commercially viable, it had many excellent properties and subsequent developments have now achieved world wide usage.

The protein fibres, or collagen, of real leather are more tightly matted near the upper or 'grain' surface than on the flesh side. This structure gives great strength, together with permeability to water vapour and air, and the ability to flex without much wrinkling of the grain surface. Certain plastic foams show similar properties but absorb moisture less readily. Felts or 'non-woven' fabrics have good strength properties but do not readily accept lacquers or

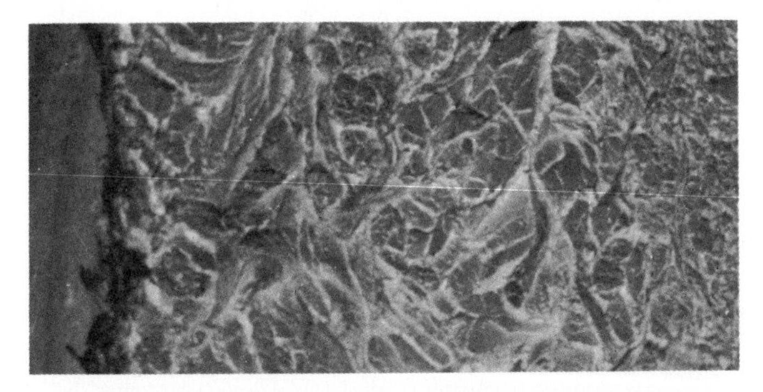

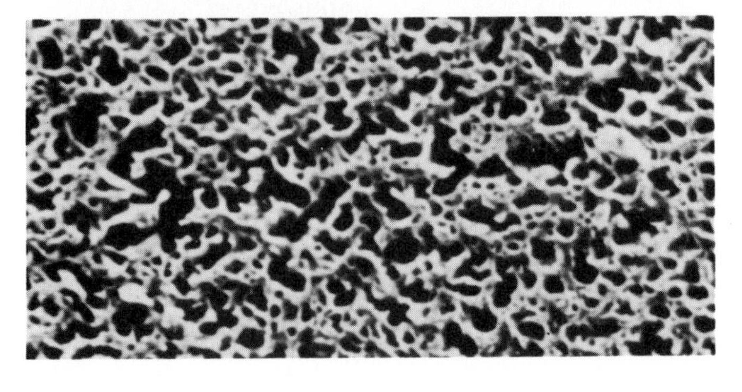

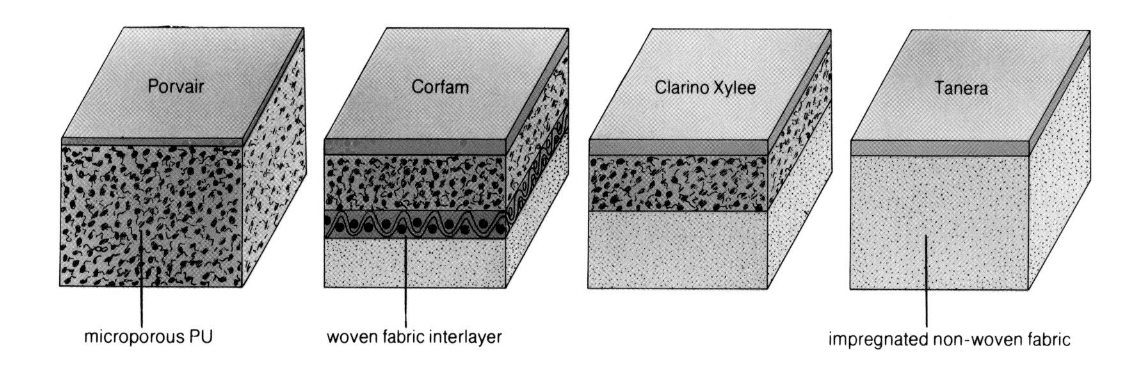

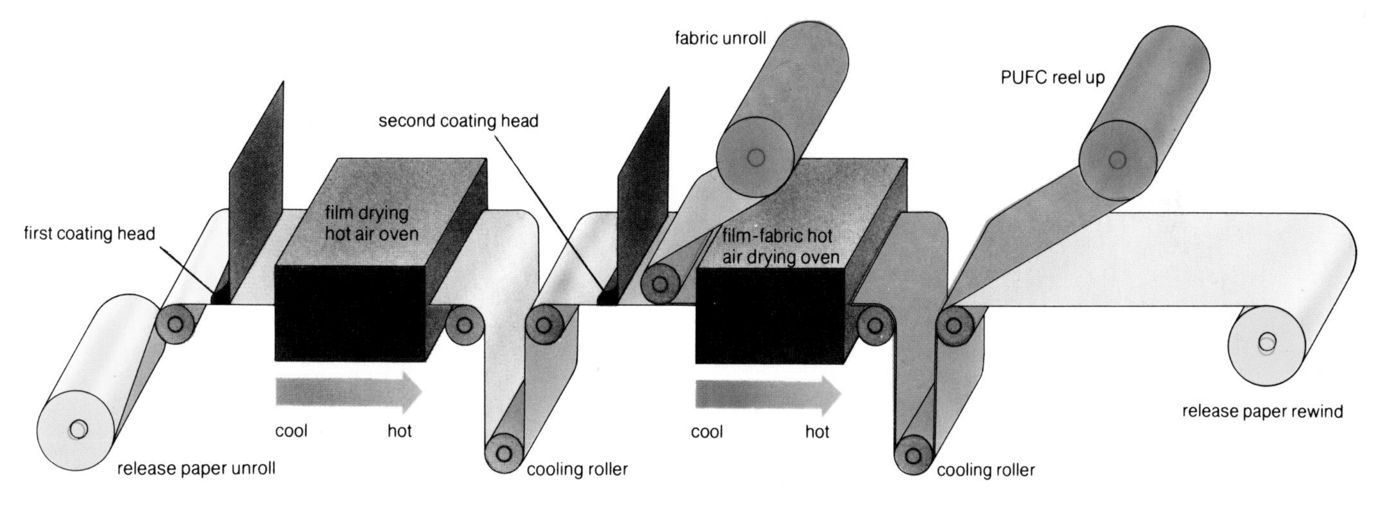

Top: various constructions are used for modern synthetic leathers, but all are coated with polyurethane, shown here in blue. Below this there may be more polyurethane in foam form (mid brown), and woven (dark brown) or non-woven (light brown) fabric.

Above: diagram of the manufacture of a polyurethane coated fabric (known as PUCF in the trade). Various fabrics and non-woven felts may be coated in this way: often a texture is stamped on the plastic surface by means of a heated roller.

Left: cross sections of real leather (top), Clarino Xylee (centre) and Porvair (bottom). In each case the grain surface is towards the right of the picture, though in the case of Porvair, which is completely uniform, it makes no difference. In real leather, it can be seen how the collagen fibres are more tightly matted towards the surface. Xylee is felt-based, polymer impregnated and coated with polyurethane foam, visible near the surface. Porvair is foam throughout, with an interconnected pore structure which gives it good 'breathability'.

finishes to give a good aesthetic appearance.

Corfam had a complex structure, comprising a non-woven felt backing on to which was laminated a tightly woven nylon cloth. The top surface was polyurethane foam, which permitted the passage of water vapour and would accept lacquers and polishes. The felt layer was made from a needle-punched web of polyester fibres, which was impregnated with a polyurethane solution to bind the fibres. The woven nylon fabric was bonded to it with a polyurethane adhesive. The polyurethane foam was cast on the surface, and the desired visual effect was applied to the surface by pressure from a heated plate engraved with the desired leatherlike grain pattern ('plating'). A final lacquering operation gave the desired surface finish.

Commercially successful poromerics range from an entirely fibrous material (for instance Tanera) to an all foam structure (such as Porvair). Tanera is produced by controlling the needle punching to give a material with a greater degree of entanglement on the grain surface than on the reverse side. The fibres are then lightly impregnated with a polymer solution which acts as a binder.

Porvair, on the other hand, remains the only poromeric which does not contain fibres of any type, and was the first poromeric to be produced in Britain. It is a *reticulated* polyurethane foam, that is, it has many pores which interconnect throughout the body of the material. It is produced by dissolving the chemical constituents of the polyurethane in a suitable solvent and dispersing finely powdered salt into the reaction mixture. The resulting material is then cast in two stages on to a moving flexible wire mesh belt.

The first layer has a coarse texture, and the second, finer layer is cast on top of it. As the polyurethane 'cures' or sets, the residual solvent is removed and the salt is leached out by washing in water. This is done by passing the continuously cast sheet material over rollers which lead it through a series of water tanks at controlled temperatures to remove the salt. The material is then dried and finished by plating and lacquering in the same way as Corfam.

Clarino, a Japanese poromeric, and Xylee, from

West Germany, are both intermediate in structure between Porvair and Tanera, that is they both use a polymer impregnated felt *substrate* on to which is cast a polyurethane foam. In the case of Xylee, careful coagulation of the foam during its formation produces the low density reticulated structure necessary to give permeability to water vapour.

All these materials are able to absorb moisture and to dissipate it through the grain surface, which is particularly important in footwear applications. Only a small amount of poromeric material is used in other less demanding applications, in which the cheaper coated fabrics are often equally satisfactory.

Top: a reverse roll coater used in the manufacture of Porvair to coat the material with cleat or coloured lacquer.

Above: a roll of Porvair passing through an embossing machine which imparts the desired leatherlike appearance to its surface. The material is compressed between two rollers, one plain and one engraved with the grain pattern in negative. Any pattern is possible: even heavily marked leathers such as crocodile may be quite successfully copied.

Above right: extruding Porvair on to a moving belt.

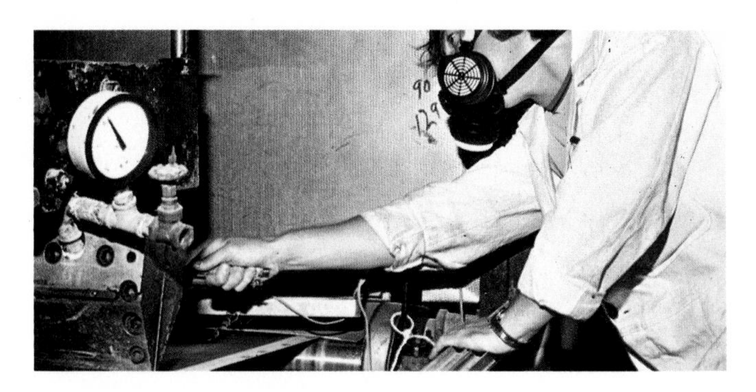

COATED FABRICS The coated fabrics can be divided into two main types, firstly those in which a polymer is coated directly onto a woven or non-woven cloth, and secondly those in which a polymer film is cast separately and then bonded to the pile of a raised fabric (or in some instances to a thin foam layer which has been laminated to a fabric substrate).

The first category are inherently simpler materials, usually made by passing a roll of fabric under a coating head. A vinyl plastisol (a solution of PVC in a solvent) is applied to the fabric by pouring the solution into a trough, the bottom of which is closed by the fabric itself. The fabric travels past a strip of metal (the 'doctor blade') at the front of the trough, which spreads the solution evenly. The fabric is then passed through an oven to drive off the solvent and allow the vinyl to cure. Any desired effects can be produced by using a hot metal plate into which the design has been etched, or by using a heated gravure roller. Finishing lacquers may be applied by spraying or roll coating.

The fabric substrates are usually conventional woven cotton materials, but needle punched non-woven felts may be used in some applications. Other fibres used include the synthetics such as rayon, nylon and polyesters.

These materials may absorb moisture but are unable to dissipate it through the grain surface, and so are better suited to applications where this property is not relevant, such as for making bags and cases.

Urethane-coated fabrics were introduced a few years ago in an attempt to simulate the appearance and handle of real leather more closely than the vinyl coated fabrics.

A pile is raised on the substrate surface (usually sateen or 'twill') by passing it under a rotating cylinder whose surface is covered with short needles, a process called *carding*. The pile is sheared to uniform thickness before coating. The polyurethane film is cast on to a sheet of 'release' paper (paper treated with silicon compounds to prevent sticking) and then coated with a hot urethane solvent adhesive. The film and fabric are brought together with the pile in contact with the adhesive, and the two are then passed through a pair of spring loaded rollers to ensure good adhesion. The solvent from the adhesive is dried off in an oven and after cooling the material may be finished by lacquering.

The urethane film is only about 0.003 inch (approximately 0.1 mm) thick, as opposed to the direct coating of PVC in the previous example which is upwards of 0.020 inch (0.5 mm) thick. It is also separated from the stable base fabric by the relatively mobile raised pile. When the material is flexed, the surface behaves much more like that of a fine grain leather than the equivalent direct coated vinyl materials.

It therefore has considerable appeal for use in women's fashion footwear and in upholstery applications, being both visually appealing and considerably cheaper than leather. It is not sufficiently durable to be widely used for men's footwear. Materials are, therefore, being produced in which a much stronger non-woven felt substrate is used and the raised pile is replaced by a polyurethane or even a vinyl foam to give some degree of surface mobility in the finished product. Materials of this type are not yet in widespread use, but are now being introduced on an experimental basis. A wide range of synthetic leather materials is now available, and many more are being developed to meet specific requirements.

LENS grinding

Methods of producing optical components, such as lenses, mirrors and prisms, have hardly changed in principle since the seventeenth century. The main requirements are that the optical surface should be polished to a highly accurate *figure*—part of a sphere for lenses and some mirrors, flat for plane mirrors and prisms. The accuracy required is extremely high by normal standards, since the surfaces have to be smoother than the size of waves they are to reflect or refract.

These measurements are so small that they are generally referred to in terms of the wavelength of light, usually the yellow sodium or green mercury spectral lines from discharge tubes, with wavelengths of about 5000 angstroms, or 0.5 micron. Thus a surface with a half-wave error would be true to one hundred thousandth of an inch (0.25 micron). In the mirrors of astronomical telescopes, the main mirror may conform to its theoretical curve (a non-spherical figure called a paraboloid) to within a millionth of an inch (0.025 micron). Because reflection bends light through twice the angle of incidence, mirrors need finer surfaces than lenses.

The principle which has always been used to produce both curved and flat surfaces is that if two discs of material are rubbed together with abrasive between them, their irregularities will eventually be smoothed out. If the rubbing motion is back and forth, it might be thought that the result would be a flat surface on both discs. In fact, both become curved, the lower stationary one becoming convex (bulging upwards) and the upper moving one becoming concave (with a hollowed out centre). To ensure a completely symmetrical surface, the lower disc is usually rotated while the upper one is both moved across it and rotated in the opposite direction. A machine can perform both these operations automatically.

The curve which results from this will be very nearly spherical in form. Most optical work relies on this principle even if the components are to be made deliberately non-spherical (*aspheric*) later on. A flat surface is regarded for optical purposes as being part of a sphere with an infinitely large radius: in practice the surface against which the item is to be ground is

Below left: a modern diamond miller designed for lenses up to 130mm (5in.) diameter. The cutting head is canted to alter the radius. It made this lens in 5 minutes, stopping automatically.

Below: hand finishing a prism. The pitch tool is kept flat by rubbing on a flat 'keeper' (front).

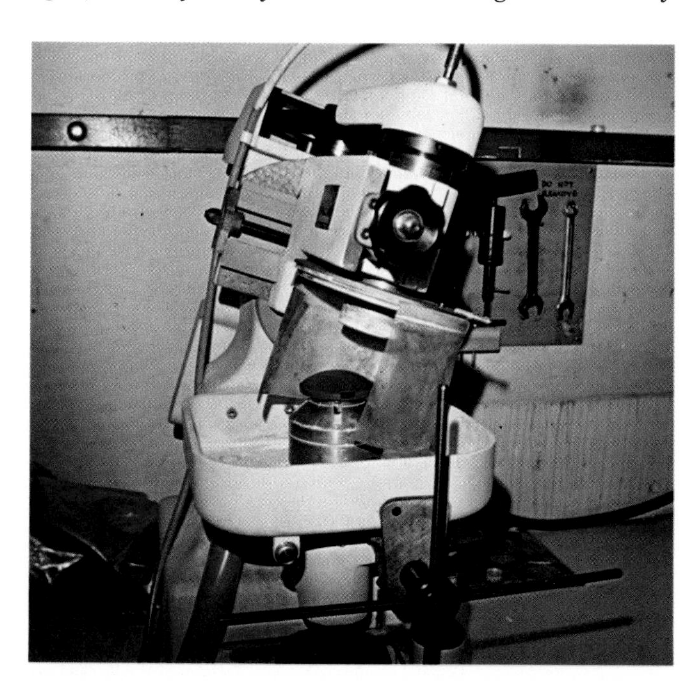

made larger than the item itself. Spherical mirror and lens surfaces are defined by their radius of curvature—the radius of the sphere they are part of. This determines how strongly diverging or converging the component is—usually steeper curves are more difficult to make.

CUTTING THE GLASS The raw glass is sawn to outline with a smooth circular saw (or bandsaw) impregnated with diamonds, cooled by a water-soluble oil, which cuts glass very quickly. Although a delicate material, glass can be machined almost like metal using diamond tools. It can be turned in a lathe, have holes drilled in it or rough figuring generated on it with a milling machine. Before such machines were available, optical workers had to grind the glass by hand, using coarse abrasives.

Components thus machined need only one or two successively finer abrasives (usually aluminium oxide powder or garnet with water) to produce a surface smooth enough for polishing. This smoothing operation can be done by machine with a number of components stuck (blocked) with wax or pitch to a jig of the right curve. This block is then worked on an iron tool which has been previously trued up on a *keeper* of opposite form (that is a convex tool has a concave keeper and vice versa). The resulting spherical curve can be checked by optical means or by a spherometer, an instrument which mechanically measures the depth of the curve across the lens.

POLISHING Precision polishing is usually done on a preformed polisher made of pitch (refined from wood or coal) using cerium oxide as a polishing agent. Low quality components may be polished on felt pads instead of pitch, while some optics require special polishing agents. Pitch is a very viscous liquid which conforms slowly to any components worked on it. Some of the skill of optical work is concerned with choosing the correct hardness of pitch because the flowing action of the polisher, changing shape as the work progresses, is critical to producing a good shape.

Other problems are flexure of the glass and the heat produced by friction during the work. The glass cools more readily at the edge than the centre and this introduces an important variable in high quality work. This has been partly alleviated by the development of types of glass and ceramics having a zero coefficient of expansion over a wide range of temperatures. This is only suitable for reflecting components, rather than lenses, owing to its crystalline structure.

After testing, the components are unblocked and the reverse sides smoothed and polished if required. Prisms, which have flat faces, are held in metal jigs or plaster blocks in a similar way. Prism angles are checked optically in an angle dekkor, which compares the job with a master prism of known angle. Correction of the angle is often made by hand (by putting more pressure on one end or the other) and accuracies of half a second of arc can be achieved.

Larger optics (say, above 4 inch—10 cm—diameter), highly curved or best quality components have to be worked singly, which makes them relatively expensive. Aspherics of any quality are made partly by machine but finished by hand. Many such curves can now be moulded, if large quantities are involved, or even turned with a single point diamond (the point of which may be only a molecule in diameter) on a specially developed lathe. Results have been claimed to one-half wavelength but this is not yet a commercial reality.

There are a number of ways in which the various types of optical components can be tested. The most common is to place the surface under test against a reference surface of the shape and accuracy required. As long as the two are approximately the same, *Newton's rings* will be seen—light and dark fringes caused by interference.

Interference is the incidence of two waves in the same place at the same time. If they are radio waves, the result is the familiar whistling noise caused by stations 'interfering'. With light waves, the light and dark fringes are produced in a similar way to the coloured rings seen on a wet oily road, and are due to the thin wedge of air between the two surfaces. If the two optical surfaces are exactly the same, no fringes will be seen. To aid visibility, it is usual to examine the fringes of light of one colour only, produced by a sodium or mercury lamp.

For non-standard surfaces, and for producing the reference surfaces, other optical tests are used, the simplest being the Foucault test. This uses a knife edge to cut off rays coming from the component which do not go exactly to the focus required.

Left: this type of lens polishing machine is widely used to produce medium-quality lenses in quantity. A steep block of lenses is shown with the polisher alongside. If Newton's rings indicate that the block has become too steep, this can be corrected by running the machine with the block on top and the polisher underneath. When both sides are polished, the lens edges are milled separately, in such a way that both sides are completely symmetrical.

A more recent test involves the interferometry of light when comparing an image with a copy of the object, a ground glass 'scatter plate', that produced the image. This has the advantage of showing up the size of errors in the image, rather than on the optical surface.

These methods only test the image on the optical axis but multiple systems such as camera and projection lenses can now be tested for their modulation transfer function or MTF. The image of a test chart, consisting of black and white bars of different spacings, is scanned photoelectrically. The difference between the object and the image can therefore be computed for all points in the field of view of the lens.

OTHER METHODS A large proportion of the popular, low priced cameras available today have single plastic moulded lenses. The first plastic lenses were made in 1934 in Britain, and their use is now widespread. They are equivalent in quality to many mass produced glass lenses, though at rather less cost. The plastic—either acrylic or styrene—deforms slightly on setting, but this can be allowed for when making the mould.

Below: a 'tree' of cheap, mass-produced plastic lenses being removed from their mould.

Above: linoleum maturing in stoves. After about 10 days it is cut up and rolled into lengths of 27.5 metres (90ft.) for delivery.

A recently developed method for the final working of glass surfaces uses a beam of ions, produced in the same sort of way as in ion propulsion devices. Ions of a heavy gas such as argon are used at a potential of about 10 kV, with a beam as narrow as 1 mm. The beam wears down the surface at a rate of about 1 micron an hour—worthwhile only in cases where a fine finish is needed. But for most jobs the accuracies involved can only be achieved by hand and optics will remain a field in which the skilled craftsman is vital.

LINOLEUM

Linoleum is the oldest and most familiar of smooth-surfaced floor coverings. Available for more than a hundred years in a wide range of colours and patterns, it has been used as a washable and inexpensive substitute for carpeting.

Linoleum was invented about 1861 by Frederick Walton, who built the first linoleum factory at Staines, England. The original method of manufacture was to hang a piece of light cotton lining material between two points in a shed or 'stove' and pour linseed oil over it. As the linseed oil film oxidized, more oil was poured until the layer was thick enough; resins and fillers such as cork and pulverized wood were added.

This process was soon replaced with a more efficient

Right: like the previous picture, this was taken at one of the few remaining linoleum factories, belonging to Nairn Floors Ltd of Kircaldy, Scotland. A run of a pattern called Melomarble is being calendered – pressed into sheets by polished steel rollers exerting enormous pressure. The marbled pattern goes right through the linoleum, and is achieved by careful mixing of pigments.

means, still in use, of oxidizing the oil by heating it in kettles until it is so thick that it will scarcely flow. It is heated further in kettles with agitators while the resin is added. Sometimes air is blown through it while it heats; then the fillers are added, as well as pigments for colouring.

Next, the mixture is calendered (pressed between rollers into sheets) and the backing is applied. The backing was originally burlap, which is still used for heavier linoleum, but modern backing is usually felt saturated with asphalt. Then the linoleum is hung in heated rooms to harden it, which may take several weeks. (Linoleum tile is made by heating it still longer, thus making it harder.)

About 1900 Walton invented an inlaying machine, which made possible a greater variety of patterns. Modern patterns which resemble inlaying are made by applying mixes of granulated colours through a stencil; these are added under pressure before 'stoving'. Marbled patterns are produced by careful mixing of pigments; the degree of marbling is controlled by adjusting the pressure of the calender rollers.

While linoleum has long been used on kitchen and bathroom floors, it is important to keep the underside free of dampness, although special saturants have been used in the backing to preserve the felt. Since World War II, the production of linoleum has been falling off in the face of competition from newer products, such as asphalt tile and especially vinyl flooring, but it is still much used in commercial building.

MANUFACTURED BOARD

Plywood, particle board and various kinds of hardboard are ingenious and useful wood products made possible by twentieth century technology. Particle board is also called chip board or pressed wood; hardboard is called fibreboard and sometimes Masonite, which is a trade name.

Plywood is constructed of several thin layers of wood, called veneers or plies, glued together with the grains at right angles to one another, with the result that plywood is difficult to split and has far more tensile strength, as well as resistance to changes in humidity and temperature, than solid timber. Poor grades of wood can be used for the interior plies, or for all the plies in the case of plywood which is to be used where it will not be seen, as in sub-flooring. Alternatively, any kind of finishing quality wood can be used for the outside ply; finished properly, sheets of such plywood make attractive wood panelling. Despite the high strength of plywood it can be worked with ordinary woodworking tools.

Particle board makes use of waste material from timber processing and of forestry thinnings (undersized trees). These are chopped into small chips, which are coated with resin and pressed into sheets by machinery, using heat. The sheets can be used for panelling, shelving and so forth, like plywood, but without its extraordinary strength.

Hardboard is made by subjecting wood chips to high steam pressure and then suddenly releasing the pressure, which causes a fresh activation of the lignin, the natural plastic substance which binds wood fibres together. The chips are then pressed into large thin sheets on heated presses. Hardboard is used mostly for interior panelling, and is available in a wide variety of finishes simulating, for example, trowelled plaster or brickwork. One of the most familiar uses of hardboard is pegboard, which has holes in it to accept wire attachments for hanging up tools or kitchen utensils, and for displaying goods in shops.

Attempts to make such wood products were unsuccessful at first because the necessary adhesives and

Left: these chips will be steam treated to loosen the lignin, then reduced to fibre by an explosion process or by grinding, then pulped and pressed into hardboard or baked into insulating softboard.

Below: pulp being poured on to a moving wire mesh belt to make a wetlap.

other technology had not been developed. Manufactured board, as well as other technical processes, received a great impetus during both world wars when combatants, especially Germany, were short of raw materials and had to make do with substitutes. Urea formaldehyde (UF) is the most common adhesive used for making plywood and particle board, although many different formulae and additives are used to make these products more moisture proof, flame resistant, insecticidal, and so forth.

PARTICLE BOARD MANUFACTURE Wood from thinnings and wastes from industrial timber processing are cut into small chips by a rotating disc with cutting edges around its face, or by a cylindrical cutter block with knife blades set in it. The chips are crescent shaped. If the equipment has knife blades, they can be adjusted to make larger or smaller chips. The chips are then passed through a grading machine which, with a winnowing action, eliminates large chips and particles of foreign matter. (At this stage or sometimes later, magnets are also used, because if particles of metals are pressed into the board they will ruin a saw blade.)

Left: in wetlap forming, the pulp is laid on to a wire mesh belt 1220mm (48in.) wide. On this, the fibres interlock, which gives the board a reasonable degree of strength – though no kind of manufactured board is as strong as solid timber, and particle boards such as chipboard and hardboard are less strong than kinds made from continuous sections of timber, such as plywood, blockboard and laminboard. This picture shows a smooth surface being put on the board by laying a finer top coating of sawdust on it. The process will be completed by heat and pressure to bond the soggy mass into a dense board, followed by cooling and maturing to even out stresses.

Regardless of source, chips contain a moisture content which must be reduced to two or three per cent. The chips are dried by being tumbled in heated drums. They are then conveyed to the tops of mixing cylinders through which they are allowed to fall; nozzles in the cylinder walls spray them with UF resin as they fall. (They still feel dry because the resin content does not exceed ten per cent.) Alternatively, the chips can be sprayed while being tumbled in drums.

Next the chips are spread on a continuous band or a line of linked steel plates. They used to be spread by hand, but this resulted in uneven distribution and subsequent weak places in the finished board. Some types of machinery can spread the chips so that they fall with the smaller chips on the top and bottom and the large chips in the middle; this makes a smoother finish on the board and a more uniform density.

The layer of chips on the plate is called a mattress. At this point the chips have a flow characteristic similar to that of coarse sawdust. To prevent loss on account of mechanical vibration or stray air currents, they go through a cold press which exerts a relatively light pressure to consolidate them. They are also slightly dampened to replace moisture lost by evaporation and to compensate for losses during subsequent heat pressing. Then the units are loaded into heat presses. For 15 mm boards (0.6 inch) the pressure is 1 N/mm_2 (150 psi), the temperature is 121°C (250°F), and this is applied for ten minutes.

The boards are then stacked to cool and to allow localized inner stresses to become evenly distributed. This takes several days. The boards are then trimmed and sanded by drum sanders on both sides.

An alternative process is extrusion, in which a reciprocating ram forces the chips horizontally through parallel heated metal plates, which are adjustable for thickness. Continuous lengths of thicker board can be produced this way, but the internal structure of the board is slightly different because of the longitudinal rather than vertical pressure on the chips. Extruded board has higher tensile but lower bending strength.

Particle board is made in three densities. High density board weighing 640 to 800 kg/m^3 (40 to 50 lb/ft^3) is used for high strength and stability, as in flooring. Medium density board weighing 480 to 640 kg/m^3 (30 to 40 lb/ft^3) is used for such applications as panelling, partitions, shelving and furniture manufacture. Lower density board is used for roof decking, ceilings, and core material for composite panelling. Particle board has sound insulating properties which improve with lower density. The bulk of particle board manufactured in the UK (175,000 tons in 1969) is of medium to high density in the range 12 to 18 mm ($\frac{1}{2}$ to $\frac{3}{4}$ inch) thick.

PLYWOOD MANUFACTURE Trees are chosen and inspected carefully before manufacture to ensure that they are used to best advantage. Manufacturers have to import wood from all over the world; trees with long cylindrical clear trunks which can be easily peeled are necessary, and many of these grow in tropical areas.

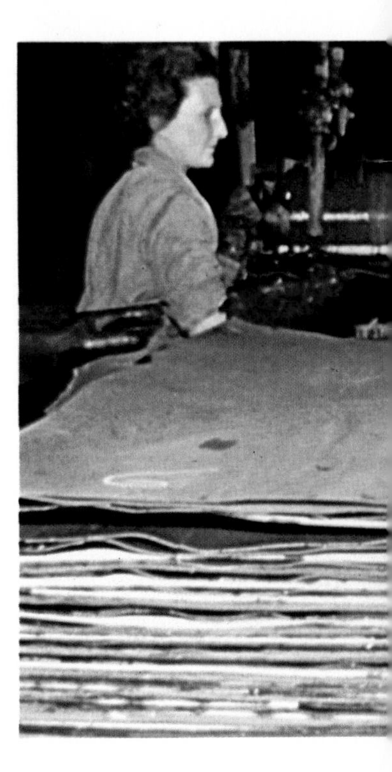

Right: plywood veneers coming out of the drier. Good-quality plywood needs careful attention paid to achieving an absolutely even moisture content throughout all the veneers; otherwise, subsequent uneven drying will cause them to shrink by different amounts and the plywood will warp, twist and split. Cheaper plywood is made with less care and does not last well as a result. Other factors in assessing the quality of plywood are the presence of any joins in the surface veneer and the number of visible knots, which weaken the board as well as spoiling its appearance; this is particularly important with types with a decorative face.

First the logs are boiled or steamed in large vats. This softens them, ensures that moisture is evenly distributed, and reduces the likelihood of splitting or tearing of the veneers during subsequent processing. Tropical hardwood especially requires this treatment; some species (such as birch and beech) can be peeled without it. In Finland, the practice is to leave the logs in the log pond for months, injecting steam into the pond to keep it from freezing.

The logs are then peeled of bark and cut to convenient lengths. The peeling of the veneers from the logs is carried out on a large lathe which rotates the logs against a knife blade running the full length of the log. Strict quality control begins at this point, for the veneers obtained must be of uniform thickness throughout. The rate of feed and the angle of the blade are adjustable for individual logs and species. Veneers are cut from about 1 mm to 4 mm (0.03 to 0.16 inch) thick. The veneer is then clipped automatically or manually to predetermined widths in such a way that defects are clipped off.

Next the veneers go through continuous tunnel driers to ensure that they are of uniform moisture content; otherwise the finished product will warp or twist. The permissible content varies from 5 to 14%, depending on the type of adhesive to be employed. (There are also wet and semi-dry cementing processes which do not require this careful drying; these result in a cheaper but poorer grade of plywood which is used where cost rather than finish is the consideration.)

The dried veneers are then sorted and graded. Veneers for faces of the most expensive grades of plywood are often full width, but joins are permissible. Joins must be smooth and parallel; this is accomplished by

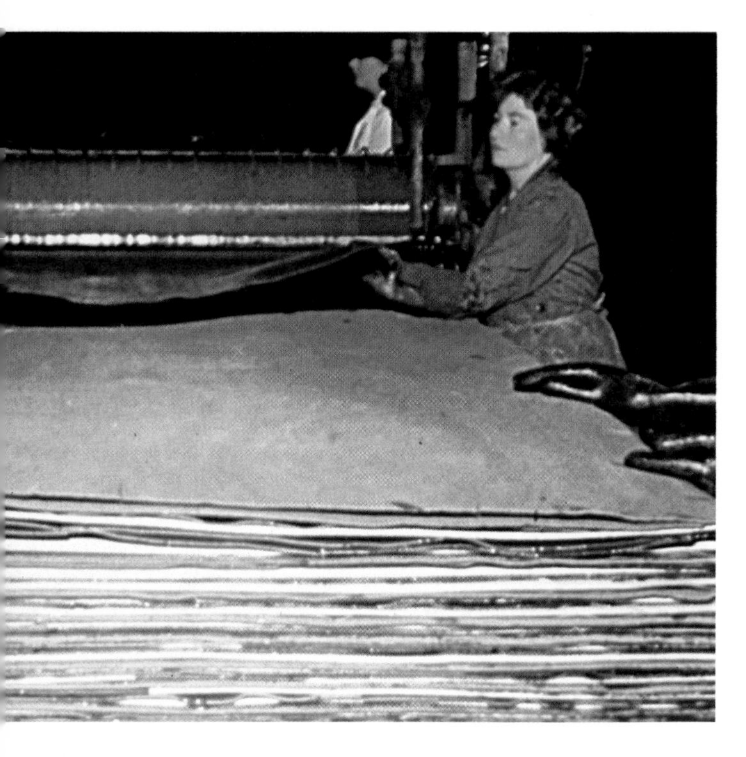

guillotines or edge joiners. Materials used for core veneers may be edge joined using staples (which are removed later) to prevent core gaps which would lower the strength of the finished board. Veneers for face joins are carefully selected for colour and are joined by an automatic tapeless splicer, which draws the edges together and makes a join which is nearly invisible and as strong as the veneer itself.

Next the glue is applied for the sandwich construction. It must be evenly applied, for too much glue will result in a poor bond. An alternative method is the use of resin impregnated paper, which is cut to size and placed between the plies. (Three-ply construction describes itself; multi-ply comprises a face, a back and usually three or more inner plies. The plies are arranged with the grain of each at a right angle to that of the plies on either side; nearly all plywood has an odd number of plies, but if the number is even, the centre two plies can have the grain following the same direction.)

The veneers are sometimes pre-pressed cold. The hot pressing is then carried out in hydraulic machinery between multiple heated platens. Temperature, pressure and so forth are adjusted to the type of construction under production. The finished boards are then stacked to allow stresses and uneven moisture content to work out, and trimmed and sanded.

Blockboard, laminboard and battenboard are constructions of face and back veneers with cores made of strips of wood placed together face to face. Block or ply constructions can be made with a saving in weight by using wood of lower density for the core material, such as pine. They can be made to order with a core material of cork, foam or fibre for applications where sound or heat insulation is important, or can be covered with plastic or metal sheeting for other special purposes. Plastic faced plywood is used for building forms for pre-cast concrete units, because it can be re-used several times and leaves a smooth finish on the concrete. Plywood can be veneered and pre-finished in several ways; choosing decorative veneers is an art in itself, because veneers from the same batch or even from the same tree may not be a perfect match. Research continues on ways to weatherproof plywood for exterior use. Moisture-proofing is not perfect yet because trimming of plywood sheets will leave the edge liable to penetration; nevertheless plywood is one of the most durable and versatile of construction materials.

MAPMAKING

The task of the cartographer (map-maker) is to represent the topographical (natural and artificial) features of the Earth's surface at a greatly reduced scale in a convenient form, usually on flat sheets of paper. His first and fundamental difficulty arises from the curvature and irregularity of the Earth's surface, and although the curvature can be ignored for maps of

Above: a 16th-century German engraving of a surveyor at work with various primitive devices for measuring angles and elevations; note the plumb line used for getting an accurate vertical. The horse would have to stay remarkably still.

Right: digitizing the information on a map – turning it into a series of numerical co-ordinates which are fed into a computer. The position of the cursor is sensed electro-magnetically, and the features are encoded with the buttons on the cursor. Such a store of information greatly simplifies the task of producing maps of the same area at different scales or showing different features, and of revising existing maps.

small areas, the surface irregularities—hills and valleys—make it necessary to show the ground features as plan projections on a plane (flat) surface. Thus, unless the ground is level, the distances shown on the map do not agree exactly with those on the surface. With larger areas, such as Great Britain, the curvature has to be taken into account and the plan projection is made on a regular curved surface conforming as nearly as possible to the shape of the Earth; this is known as the spheroid of reference.

Maps on curved surfaces, however, are inconvenient objects to handle, and a transformation has to be made from the curved to a plane surface by means of a map projection. There are many different map projections, but they all result in the distortion, in one way or another, of the pattern of features on the curved surface. This distortion is very obvious in atlas maps of the whole world (as on Mercator's or Mollweide's projections), but it is possible, by using particular projections, to retain some elements of correctness.

Orthomorphic projections are widely used for large scale topographical maps. In an orthomorphic projection the scale does not remain constant over the whole area, but at every point the scale is the same in all directions for a short distance, thus preserving the correct shape in small areas. The orthomorphic Transverse Mercator projection is used for the official maps of Great Britain, and in this comparatively small area, the scale variation caused by the projection does not exceed one part in 2500, which is negligible for most purposes.

PRIMARY MAPPING A distinction must be made between the two classes of mapping. Primary mapping is produced directly from a topographical survey; derived mapping at smaller scales results from the reduction and generalizing of primary maps, or from the compilation of map material drawn from either or both categories.

When undertaking primary mapping, it is not possible to fit together individual surveys of small areas in order to complete the survey of a large area. The errors in such independent local surveys would accumulate and produce discrepancies that would be unresolvable. A consistent framework has first to be constructed covering the entire area to be mapped. This framework must be located and oriented correctly on the Earth's surface, and this is accomplished by astronomical observations at several points within it.

Since the middle of the eighteenth century the classical control framework for mapping large areas has been produced by triangulation, in which the angles of a system of triangles are measured to a high degree of precision with a theodolite (an instrument consisting basically of a telescope moving around a circular scale graduated in degrees). The linear dimensions are determined by base measurement; that is, by measuring the lengths of several sides in different parts of the system, and thereafter by calculation. Accurate base measurement was first carried out in Great Britain with glass measuring rods on Hounslow Heath in 1784. At about the same time great improvements were made in the design and manufacture of theodolites, particularly by Jesse Ramsden (1735-1800), and the national triangulation of Great Britain was begun. The lengthy and laborious operation of base measurement with glass rods (1784), bimetallic bars (1826), and later with steel tapes suspended in catenary (the curve a string makes suspended between two supports), was gradually replaced from the 1950s onwards by electromagnetic methods.

Instruments such as the tellurometer and Geodimeter, which measure distance by recording the time taken for an electro-magnetic wave to travel to and from a remote station, enable the whole operation to be completed in a few hours and make it possible to incorporate many more linear measurements into the triangulation framework. The subdivision of the main or primary triangles into secondary and tertiary triangles is continued until a density of fixed control points is obtained, which meets the requirements of the topographical survey method chosen.

A similar control network for heights above the sea level datum is obtained by a pattern of intersecting

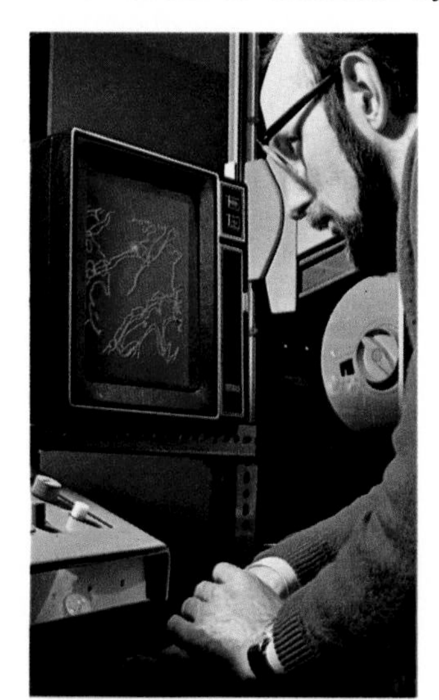

Left: checking an edited map on a computer display screen.

Right: triangulation. A series of triangles is built up from a set of base triangles; a second series is used as a check. In the geometric model method, pairs of aerial photographs are viewed together to produce a 'model' from which distances and heights are taken for the preparation of the final map, complete with contour lines.

lines of spirit levelling. In Great Britain evidence of the levelling may be seen in the broad arrow benchmarks (fixed points of reference used for levelling in surveying) on permanent objects and buildings. The height of each benchmark is shown on the large scale national maps.

AIR SURVEYS The topographical survey can be undertaken on the ground or by means of aerial photographs. The ground surveys of the eighteenth and nineteenth centuries were carried out with compass, plane table (essentially a drawing board mounted on a tripod together with an alidade consisting of a rule with sights at both ends to give direction of survey points from the table) and chain; in the second half of the twentieth century, new instruments which combine the theodolite with either an optical or an electromagnetic distance measuring device came widely into use. The method most commonly practiced, however, in recent years for topographical mapping is air survey, because of its speed and economy.

Except when the ground is absolutely flat, and when the photographic exposure is made with the camera pointing vertically downwards, an air photograph cannot be used directly to make a map because of the scale variation caused by the ground relief and the tilt of the camera. To solve the problem, stereo plotting machines, such as the Wild photogrammetric machine, are generally used. Pairs of overlapping air photographs are set up in the machine in accordance with the control data already obtained, so that their positions and orientations in space at the two moments of exposure are recreated. It is then possible to create in the machine a three dimensional 'model' of the ground from which planimetric detail (distances and positions of features), contours and spot heights can be derived. The 'model' is formed by the intersection of the images of the two photographs.

In primary mapping, the task of the cartographer is to show all the information collected by the surveyor in the way that enables it to be most clearly and readily comprehended. With primary mapping at large scales (1:500 to 1:10,000) the preservation of positional accuracy is of first importance, and the cartographic process of generalization has little part to play.

PRINTING One method commonly used in drawing offices for making a single colour primary map is to scribe the detail on plastic sheets. A clear plastic sheet is covered with a semi-opaque waxy coating on which the image of the surveyor's work is printed by a photographic process. The draughtsman cuts away the waxy coating along the lines of the surveyor's drawing, leaving very sharply defined lines of clear plastic. This operation produces a negative from which a positive on another sheet of film or plastic can be made by contact photography. Names and symbols are usually added at this stage; these are printed on strips of very thin film which are stuck to the film positive. A lithographic printing plate is made, also by contact photography, for printing on a rotary offset litho machine.

The combination of scribing and photographic processes compares very favourably for speed and quite well for quality with the copper plate engraving used in the early nineteenth century for map re-

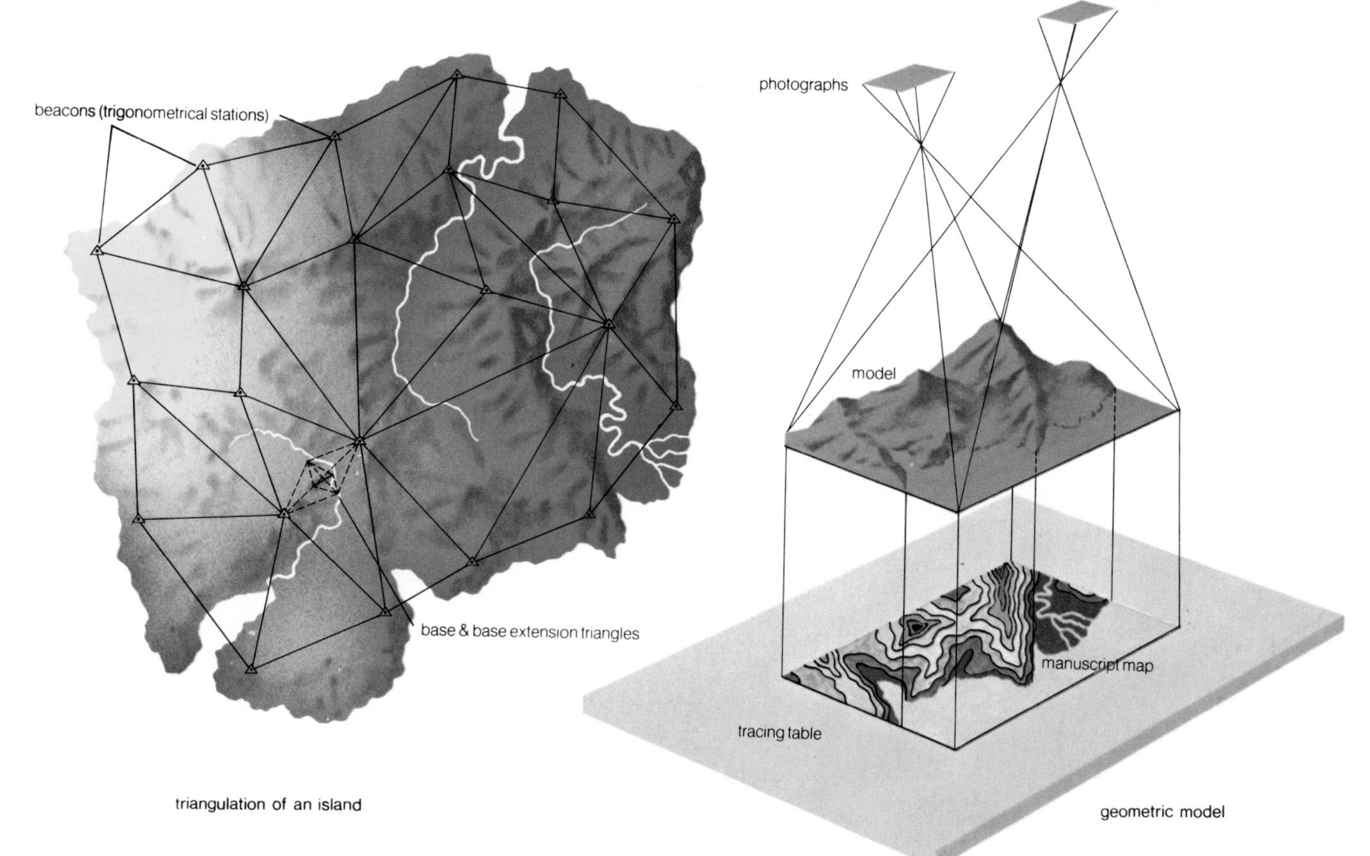

triangulation of an island

geometric model

production. In this process a hand-drawn tracing was made of the surveyor's drawing and this was transferred to the copper plate which was coated with wax to take the transfer. The engraver then cut along the transferred lines into the copper; the outline, names, symbols and ornamental drawing, including the hachures (lines used to shade a plan) representing the hills, were all engraved in this way. Ink was then rubbed into the engraved lines and, after the surface of the plate had been cleaned, impressions were taken on paper. Revision was carried out by hammering the copper flat in the area to be revised, and then engraving the new detail.

The revision of primary mapping in a highly developed country depends upon a flow of new information from surveyors who may be permanently occupied in surveying changes as they occur. In the drawing office it is important that the new work should be incorporated into the map without having to redraw the whole, as this would lead to a progressive deterioration in accuracy. The combination of new and old is therefore achieved by photographic methods.

DERIVED MAPPING Small scale derived maps are made either directly from the primary mapping or from other derived maps. Here there is great scope for the cartographic designer because it is generally necessary to make a selection of detail for showing at the smaller scale, to generalize the outline of some features, and to exaggerate the size of others so that they are given the prominence the designer requires. Because of the closeness of detail on small scale maps, it is usual today to use colours to distinguish one type of feature from another, such as red for main roads, blue for water, and green for woods.

A typical procedure for making a smaller scale derived map from primary mapping is as follows. A print of the primary map is made in a 'non-photographic' blue colour (which will not show up in a photograph), either at the scale of the primary map or at a scale intermediate between it and the derived map. The features for the derived map are selected, generalized and penned in black on the blue print, which is then photographed and reduced to the final scale. Several blue prints (depending on the number of colours) are made of the reduced drawing, and separate drawings in black ink are made for each colour.

A printing plate for each colour can then be made from the separate drawings, and, after printing, the colours will all be in correct register. Although each colour is printed separately, it is possible to produce several shades of the same colour by means of rulings or stipple. The relief of the ground, which in the nineteenth century was generally shown by hachures, is now normally depicted by contour lines, which may be enhanced by hill shading or by layer colours. In the latter method, the spaces between the contours are filled with a progressive series of tints, often with green for the lowest level, and deepening shades of buff and brown for the higher levels.

Maps of sea areas, made for the guidance of mariners, are known as hydrographic charts. They show coastlines and coastal features, navigational lights and marks, soundings and underwater contours. Because of the changing character of the sea bed in coastal areas, up to date information is essential, and careful arrangements must be made to disseminate revisions.

COMPUTER DRAFTING The application of computers and automatic drafting machines to mapping has been extensively studied since the late 1960s. The concept of a computer-based topographical data bank containing classified and coded information, any selection of which could be plotted at a wide range of scales, has great interest both for the makers of atlas maps and for those responsible for topographical map series. The lines representing the features of a map can be stored in coded numerical (digital) form on magnetic tape. For example, a straight line can be represented by the map coordinates of its two ends; curved lines by a series of such coordinates.

Trials have shown that the information on large scale maps can be stored in digital form and then drawn automatically by high speed drafting machines. The same information, or selected categories from it, can be drawn automatically at smaller scales, and the possibility of producing a whole family of derived scales in this way is very attractive to map makers. The generalization required, however, for small scale maps, such as the widening of roads, the simplification of outlines, and the adjustment of adjacent features to conform with these changes, presents programming difficulties.

Left: a stereoplotter used to prepare large-scale maps from series of overlapping aerial photographs. Viewed stereoscopically, any pair of photographs of the same point taken from two slightly different places will produce a correctly proportioned, 3-D image.

Below: preparing a negative of a map.

MATCHES

The invention of the friction match in the early 1800s was described by Herbert Spencer, the philosopher, as 'the greatest boon and blessing that had come to mankind in the nineteenth century'. Although methods of generating fire by spontaneous chemical reaction had been known since the isolation of white phosphorus in the second half of the seventeenth century, friction matches were not produced commercially until 1827, by John Walker in England. Walker's matches were of the 'strike anywhere' type being made from potassium chlorate, antimony sulphide, and gum arabic. They were, however, somewhat unreliable and difficult to strike. Subsequently, and up to the end of the nineteenth century, strike anywhere matches almost invariably contained white phosphorus. The incidence of 'phossy jaw'—a particularly horrible industrial disease—among workers in match factories rendered it imperative to find a substitute for this material and, from the turn of the century, phosphorus sesquisulphide has replaced white phosphorus in strike anywhere matches.

THE SAFETY MATCH The discovery in the 1840s of a much less reactive form, or allotrope, of phosphorus, called red phosphorus, provided a material which opened the way for the production of safety matches, which are struck only on a prepared surface. The striking surface of safety matches contains red phosphorus bound to the side of the box with gum arabic, urea formaldehyde or other powerful adhesives. The potassium chlorate contained in the match head is, on striking the match, brought into contact with the phosphorus on the box and the resultant chemical reaction generates sufficient localized heat to initiate the burning of the match. Safety matches, in addition to potassium chlorate, contain sulphur with which it reacts, and diluents, such as powdered glass and iron oxide, which are inert and so control the burning rate. A binder is also required; this is normally animal glue, which consolidates the constituents and binds them to the end of the match stick.

Strike anywhere matches contain phosphorus sesquisulphide, which is more reactive than sulphur and, in consequence, friction on a rough surface is sufficient to trigger off a reaction between this material and potassium chlorate, which causes the matches to ignite.

Colour matches of either the safety or strike anywhere variety are made by the addition of suitable dyes and, in these instances, highly coloured ingredients, such as iron oxide and manganese dioxide, are omitted and replaced by a white material which is normally zinc oxide.

THE MATCHSTICK After the various chemicals, wood is probably the most important material used in match making. Various other materials can, however, be used for the sticks of matches, including wax coated cotton, paper and cardboard. Timber for matchsticks should be white and odourless, straight

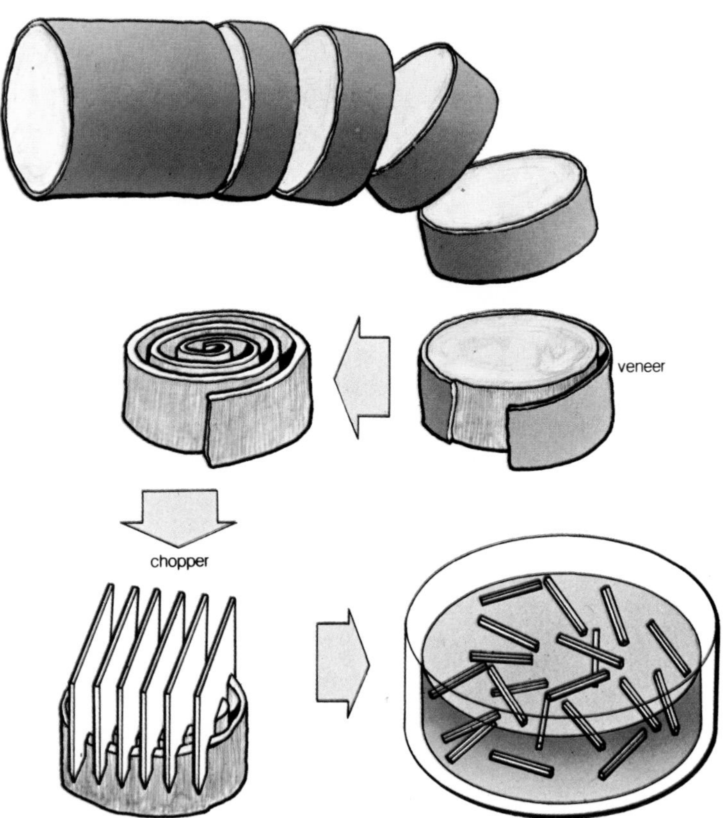

Above: a 19th-century English match factory. At this time match factories were notorious for the hardship of their working conditions, as well as for the appalling industrial disease 'phossy jaw'.

Left: the diagrams on this and the next page show the match production process. First logs are cut into billets, peeled into veneers, cut into splints and impregnated with ammonium phosphate.

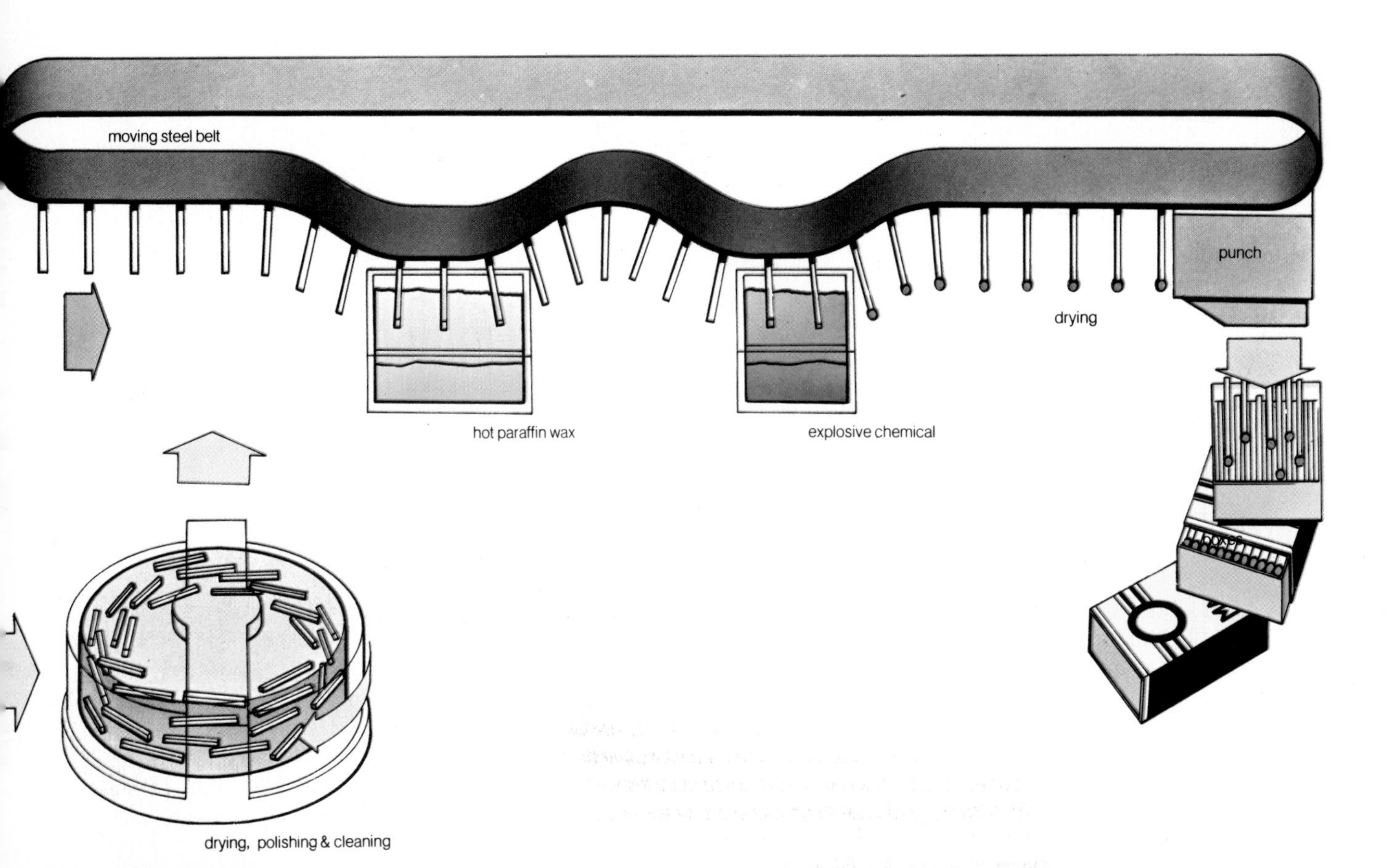

Above: the match production process continues with drying, polishing and cleaning the splints in a rotating drum and attaching them to a steel conveyer belt. The heads are then dipped into melted paraffin wax and afterwards into the head composition (which varies, depending on what kind of matches they are). After slow and careful drying they are punched out of the belt and boxed. Book matches that are made out of cardboard are prepared in a similar way except for the very beginning of the process.

Left: freshly cut splints, that is, unheaded matchsticks, are blown along steel tubes and dropped on to a conveyer belt which takes them to the match machine, where they fall through spaced holes into the machine's slowly moving steel belt.

Above right: the match machine, showing the zigzags in the belt to prolong drying time.

grained, easy to work and sufficiently porous to absorb paraffin wax. It must neither be too hard nor too soft. If it is too hard it will not absorb paraffin wax; if too soft, it will bend out of shape. The ends of the matchsticks which carry the heads are soaked in hot paraffin wax before the head is put on. This helps the match to burn by transferring the flame from the match head composition to the stick. Without the paraffin wax the match would go out as soon as the combustible matter in the match head was burnt. Matchsticks are also impregnated with ammonium phosphate, a fire retarding agent, so that they will not continue to glow or smoulder after being put out.

To make the matchsticks, logs weighing about 125 kg (276 lb) and about three metres (9.8 ft) long are first sawn into billets about $\frac{1}{2}$ metre (1.6 ft) long. These are

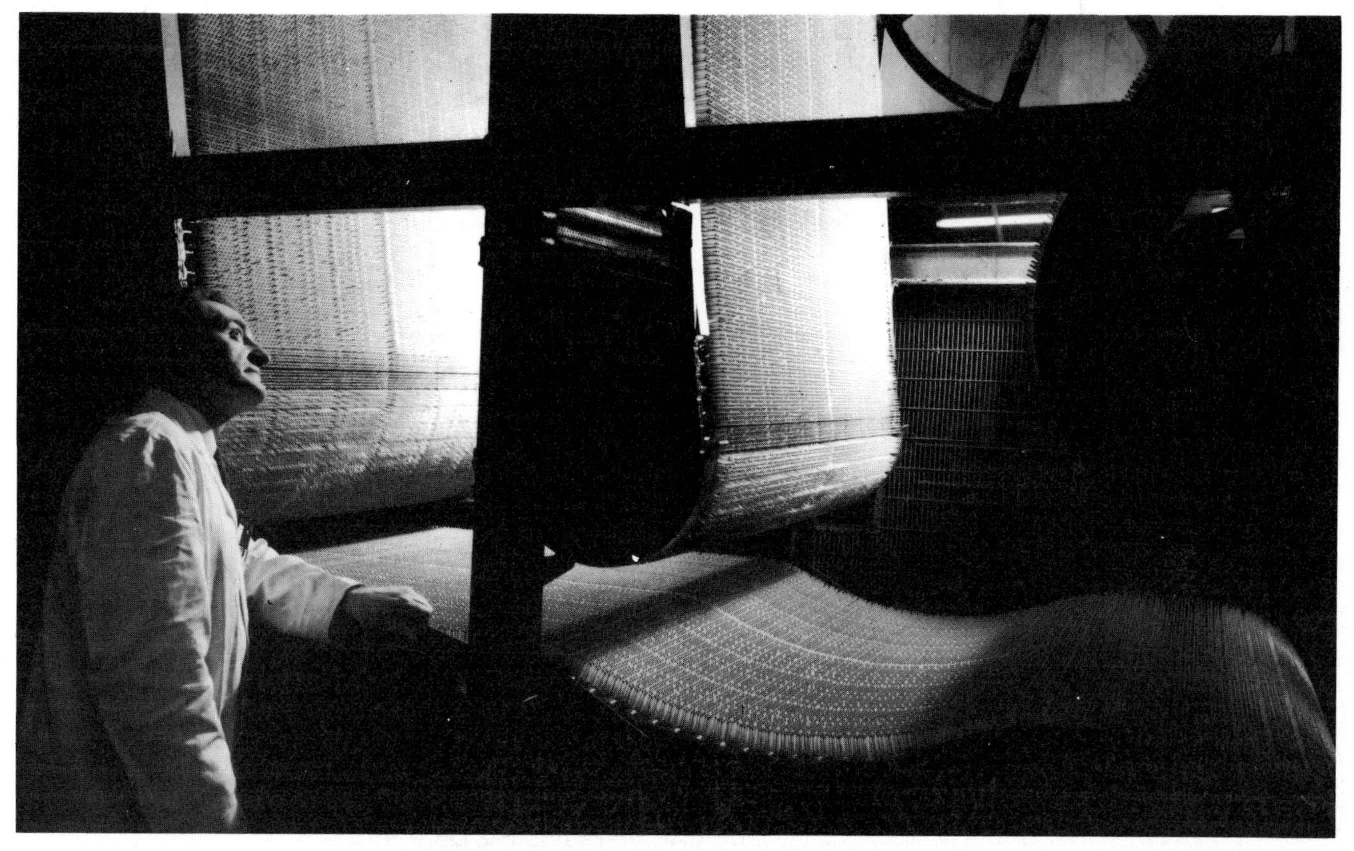

cut into veneers on a peeling machine. The long strips of veneer, about 2.3 mm (0.09 inch) thick, then move on to the choppers. Every cut of the chopper through a stack of veneer produces about 1000 matchsticks, or over one million in less than five minutes. The sticks are then impregnated with ammonium phosphate, dried, polished and cleaned in rotating drums. Finally they are blown along steel pipes to the match machines.

THE CONTINUOUS MATCH MACHINE These machines can be operated by as few as eight people and can produce and box about 20 million matches per day. The sticks are fed automatically into suitably spaced holes in the machine's slowly moving steel belt. First they pass through a bath of hot paraffin wax. A little further on the steel belt stops over a table covered with the chemical composition for the match heads. The sticks, nearly 6000 at a time, are lowered into the liquid mixture and as they are raised each stick carries a drop of composition which forms the striking head. The next stage is to dry the head, and this must be done very slowly or the match will not strike properly. It takes 50 to 60 minutes of slow movement through the machine before the matches are ready for the boxes. At any time during the day there are about two million matches on the machine. When the drying process is finished the steel belt travels downwards to the inner boxes which are moving steadily across its path. The machine then punches the matches out of their holes in the belt so that they fall neatly and in the correct number into the inner boxes as these pass by. The filled inner boxes move on conveyors parallel to the outer boxes. Both conveyors stop momentarily and 16 inners are rapidly pushed into 16 outers, and the conveyors move on again. This operation is repeated 50 times every minute.

Book matches, which consist of a comb of wood or cardboard stapled into a cardboard cover, are made in very much the same way. The combs are first cut from a strip of wood or cardboard and then fed to the match machine. When the match heads are dry, the combs are passed to a booking machine where the covers are attached.

MOVIE FILM production

The production of a motion film is a long and complicated process involving a wide range of techniques and equipment. The screenplay which sets out the film as it will appear on the screen is broken down into shots in the shooting script, carrying instructions for the technicians including such details as whether it is to be a long, medium or close shot.

It is invariably necessary to shoot a film out of screenplay order. Scenes requiring the same location or set (in a studio) or a particularly highly paid performer will be grouped together for shooting purposes. The normal film making procedure is to use only one camera. A scene is therefore shot many times from different camera positions to create different points of view and types of shot (long shots, close ups, for example). These separate shots may overlap to varying degrees to allow a choice at the later editing stage.

There will also be different takes of each shot.

The film in the camera is normally 35 mm colour negative, giving a picture size of 0.87×0.63 inch (22×16 mm). The camera runs at 24 frames (single pictures) per second. The sound is recorded on ordinary $\frac{1}{4}$ inch (6.5 mm) magnetic tape using tape recorders usually running at $7\frac{1}{2}$ inches per second (19 cm/sec). The exact speed of the camera and tape recorder are controlled by a system of pulses which are recorded on the tape to guarantee the correct speed of the magnetic film on to which the sound is transferred later. This magnetic film has a clear base of the same dimensions and with the same sprocket holes as the picture film but is coated with a metal oxide like magnetic tape. To identify each shot and to provide a synchronization (sync) point for the lining up of sound and picture later, a clapper board is shot at the front of each take. When lining the two up ('syncing'), the picture frame where the two halves of the board meet are synchronized with the first modulation of the corresponding bang on the track.

The day's shooting, called the rushes, is viewed the following morning to check the performance and picture quality. The laboratory has developed the negative overnight and made rush prints from it. The rushes are not graded: that is, they may show variations in colour case and density.

EDITING The editor will view the film on an editing machine of the Movieola or Steenbeck type, capable of running the picture in perfect sync with the sound (now on 35 mm magnetic film) both forwards and backwards at various speeds and frame by frame. To keep picture and sound in sync during editing, they are always kept in a synchronizer (up to four sprocketed wheels in parallel on the same shaft) when cutting. The cuts are made on a tape joiner, consisting of a block for joining and a knife edge for cutting. All the joins are made with clear sticky tape so that alterations can easily be made. Gradually the 'cutting copy' is built up until the director and producer are satisfied. The soundtrack will be incomplete, consisting mainly of dialogue without music and effects. Dialogue will often be re-recorded by post-synchronization using short sections of scenes as loops. This way the artists can recreate their performance while hearing and watching the original, and a better quality recording made. Sound effects will often be specially recorded.

The sound editor will make up music and effects tracks and prepare them for the final 'dub' when all the different elements will be mixed together. He may also dub a special music and effects track (M & E track) for foreign sales. The local distributors will dub in the dialogue in the required language.

The camera negative is stored in the laboratory while the editing takes place. During this period it will be decided what optical effects are required. One scene can dissolve into another, and scenes be superimposed on each other if need be. In the same way, rain or lightning may be optically added. This is done on an optical printer, which may have the two separate shots to be combined running simultaneously,

Left: on rare occasions more than one camera is used. A spectacular staged crash, or a location car scene like this one where continuity raises problems, are examples. An alternative here would be to project the outside views on to screens outside a stationary car – not very realistic.

Below left: a more typical location arrangement. Note the camera dolly for smooth tracking shots.

Below: a Steenbeck machine for editing 16mm film, enabling one film (at rear) and two sound tracks to be run in complete synchronization, or else for the film to be shown one frame at a time. Provided that the film and sound track are kept on the six spools of the machine, synchronization will stay correct whatever happens, which is most important.

with a system of half silvered mirrors making it possible to project them both on to duplicating film. A shutter can control the brightness of each separate shot, making it possible to produce fades and cross-fades (dissolves). The same type of printer may be used for adding superimposed titles. To allow freedom to experiment in making such optical effects, duplicates are made from the original negative which can then be

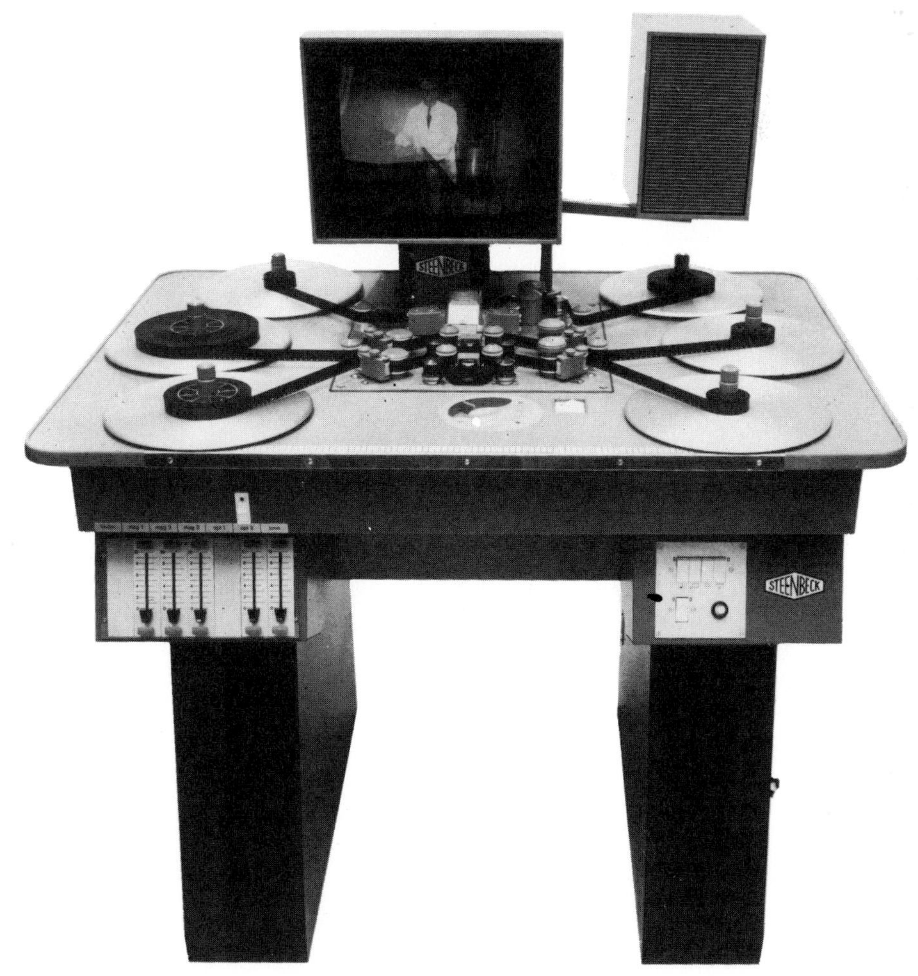

kept safe. One can make an interpositive from the original negative, and a negative from that, or alternatively, a CRI (colour reversal intermediate) can be made producing a duplicate negative in one process, at lower cost but of poorer quality.

The duplicates are made either on an optical printer projecting the image on to the duplicating stock, or on a contact printer bringing the two stocks into contact with their emulsion sides facing each other. Contact printers usually have the two stocks moving continuously over a light source (continuous printer), while optical printers can be combined continuous printers and step printers (working frame by frame).

When the cutting copy has been finalized incorporating optical effects, the negatives (original and optical effects) are cut to match it. This is simplified by key numbers exposed on to the edge of the film during manufacture and thus transferred to the cutting copy in the rush printing. The negative is spliced with film cement.

GRADING The picture negative is then viewed on a colour video analyzer by the grader, who will be able to determine the colour and density corrections to be introduced in the making of the first 'married print' with both sound and picture. After seeing this first answer print, smaller corrections will usually be made before release printing will start. To preserve the original cut negative, a graded CRI is often made and the release prints made from this.

If large bulk orders are expected it may be desirable to use the Technicolor process. This is not strictly a photographic process, but a dye transfer process using matrices in the form of 'separation' black and white prints. These are made from the colour negative, each through a different colour filter, producing a gelatin record of the respective cyan, magenta and yellow content of the original image. These matrices are then saturated with the corresponding dye and mechanically printed on to a clear gelatin coated film. Making one matrix for each of the three colours gives greater control in the final dye transfer print.

ANIMATION Cartoons and animated films are made frame by frame. Usually 12 drawings have to be prepared for each second of film, each being photographed twice. In most cases the background and some part of the character remains constant, while only a small part of the picture—head, legs or arms—is altered for every other frame. This is made easy by painting the moving part on a series of transparent overlays ('cels') which are then in turn placed in accurate register against the character's body and background, and the frame exposure is made. Complex actions such as running are drawn as cycles making it possible to use each cel repeatedly. The sound for a cartoon is usually recorded first and the soundtrack is analyzed frame by frame on a Movieola. This analysis is used by the animator to decide the number of frames to be taken for each piece of action. All the information about the artwork to be shot is listed on a 'dope sheet' which is given to the animation camera operator

A computer turns Groucho Marx into Elvis Presley. Scanning the pictures photoelectrically, as here, or using a tracing device, the starting and finishing pictures are encoded as numbers. The computer then changes the numbers in a steady sequence to alter the pictures.

together with all the artwork. Small computers have been used for some time in helping to compute the camera moves and also to control the servomotors effecting these moves.

It is also possible to use computers to help with the animation itself. The animator will only have to draw the key drawings and major in-betweens, (say, every sixth image). The computer will determine the smooth movement of each component line in the character from one position to the next in the required number of steps. This technique presupposes that the movement can be mathematically formulated.

Films can also be made using models which are moved slightly from one frame to the next. Similar techniques are used with actors to create 'impossible' special effects, such as an unlikely 'hole in one' during a comedy golf sequence. People can also be made to disappear from the middle of the scene by stopping the camera while the actor moves out of view. Any other actors must keep perfectly still until the camera starts again.

THE SOUNDTRACK In 1906 a Frenchman, Eugène Lauste, who had worked in Thomas Edison's laboratory, became interested in adding synchronized sound to the then very new motion pictures. Lauste's method was to photograph sound on to the actual picture negative. Inside the camera he fitted an exposure lamp and solenoid coil with a slit diaphragm, to which he connected a telephone microphone. Speech signals from the microphone vibrated the diaphragm assembly, which he called a light valve, causing variations in the intensity of light falling on to the film. Lauste also had to construct a special projector to reproduce this track, but since audio amplifiers had not yet been invented he was unable to reach a wide audience.

In 1922 an American scientist, Theodore Case, actually photographed sound waves by modulating an oxy-acetylene flame, but soon abandoned the commercial possibility, with his invention of a photoelectric cell. He was able to demonstrate talking films in 1923, using a gas discharge tube for recording and

Opposite page, far left: a video colour negative analyser, for grading films. A grey scale is included on the film when shooting as a colour check.

Near left: a computerized animation stand. The cels are laid on the motorized baseboard, which is controlled by an analog computer.

Above right: this optical printer is being used to superimpose a previously photographed title in colour on a moving background.

Right: 35mm soundtracks, left to right, standard magnetic stripe without pictures, but including the balance strip, usually left unrecorded, though it could be used for special effects; then full magnetic base used in the editing stages; negative optical track; and the same track, positive and with pictures, as used in the projector. Note the double track for stereo sound.

his light sensitive photocell coupled to an Audion amplifier and loudspeaker. But film laboratories at that time were unable to give consistent results, and the surface noise for the film was extremely high.

The first motion pictures exhibited with synchronized soundtracks began to appear in 1927, such as *The Jazz Singer* starring Al Jolson. The sound for these early films was recorded either photographically on the film itself, or on a separate synchronized disc, so all film projectors had to be equipped for either system. The discs were 10 inches (25 cm) in diameter, played at a speed of 33⅓ rpm, and contained a start mark on the inside groove—the end of the recording being the outside groove. There were 96 grooves to the inch, so that each disc lasted for 12 minutes, which was just sufficient to accompany 1000 feet (305 m) of picture film running at 24 frames per second. Since they were shellac pressings, these discs were extremely fragile and did not last more than a few years.

The early thirties showed a great improvement in laboratory control and film emulsions were manufactured particularly for sound on film recording. A system of background noise reduction was introduced in 1932, and most films today use photographic or optical soundtracks.

OPTICAL SOUND Cameras for producing optical soundtracks are available for 35 mm, 16 mm, and Super 8 mm film. The basic design consists of a light-tight film chamber, a film transport system with flywheel and sound drum, a detachable magazine for holding exposed and unexposed film, and an exposure lamp complete with optical system and modulator, to expose a track which varies in step with the sound vibrations. Two types of modulator are used, a mirror galvanometer and a light valve. Both produce a variable area sound negative, in which a white band in the centre of a black strip changes its width. The galvanometer reflects light from the exposure lamp, via a V-shaped mask, through a narrow horizontal slit. The light valve passes light directly through to the slit. In both cases an image of the slit is focused on to the film as it passes round the sound drum. A noise reduction bias is applied to the modulator when there is little or no sound, so that it passes the minimum amount of light, thus reducing unwanted background noise.

Negative exposure is accurately controlled, and development is carried out at a commercial laboratory on a continuous processing machine. The negative is developed to a specified density, which has been determined by previous tests with signals at two frequencies (usually 400 Hz and 6 kHz for 35 mm film). Since the negative is not suitable for reproduc-

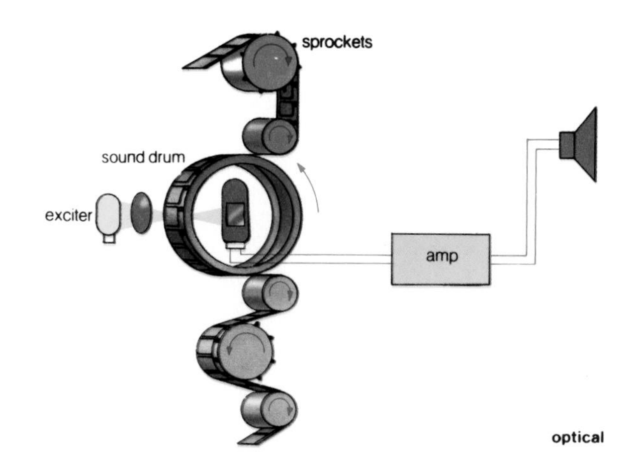

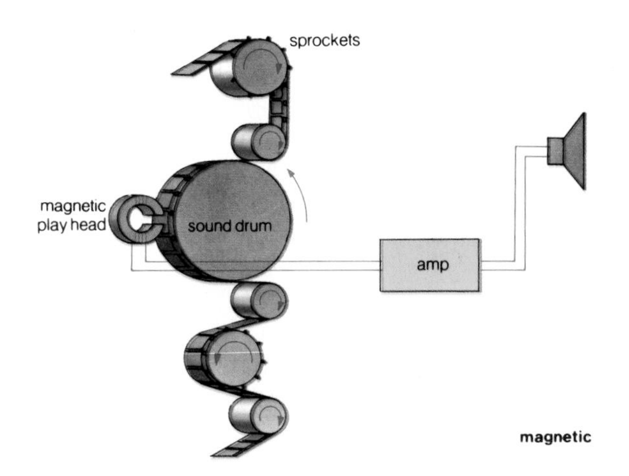

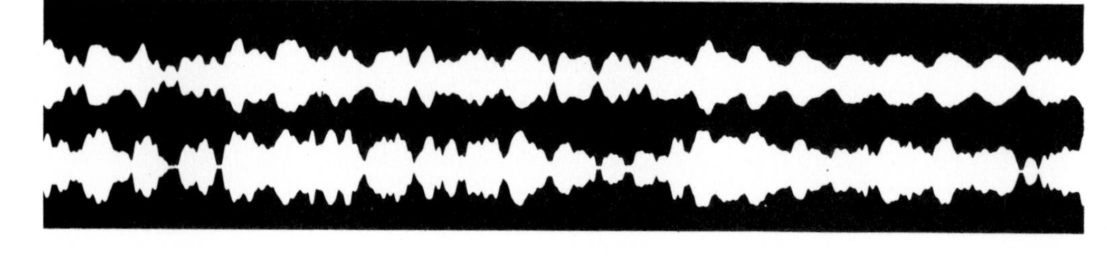

Left: an early optical soundtrack, made by Eugene Lanuste in 1910 and using basically the same system as today's. Commercial success was impossible until the arrival of electronics and loudspeakers. Below this, on a larger scale, a stereo soundtrack. This one uses the Dolby noise reduction system, invented for magnetic tape but equally useful here.

Right: machines for making soundtracks.

Left: layouts of the soundtrack playing systems for both optical and magnetic tracks on a 16mm projector. The edge of the film projects beyond the sound drum so that the light can shine through it. For a magnetic soundtrack the optical equipment is swung away and the magnetic play head moved into the same space. This works in exactly the same way as the head of an ordinary tape recorder. Since the film is required to run through the projection gate jerkily, but through the sound gate smoothly, the piece of soundtrack corresponding to a particular frame is 26 frames in advance of the picture.

tion, a print must be made before the soundtrack can be heard.

The optical soundtrack is printed on to the picture film in such a way that it is in advance of its corresponding picture frame by 20 frames on 35 mm, 26 frames on 16 mm, and 22 frames on Super 8 mm. This is because the film has to run through the projection gate jerkily, each frame of the film being motionless for a fraction of a second. The film must be running as smoothly as possible by the time the sound is picked up, so it is first passed round a series of loops, rollers and a sound drum. The sound drum is quite heavy so that its inertia smooths out any slight ripple which might occur.

The reproducer consists of an exciter lamp, slit and optical system, together with a light sensitive cell behind the film. The older type of photoelectric cell invented by De Forest has now been replaced by a photo-emissive cell or an integrated circuit chip comprising a photodiode and amplifier. Soundtracks have an overall frequency response of 50 Hz to 8 kHz on 35 mm, 80 Hz to 6 kHz on 16 mm, and 100 Hz to 4 kHz on Super 8 mm. The frequency characteristic can be altered, and the signal-to-noise ratio improved, by the use of the Dolby noise reduction system.

MAGNETIC SOUND Magnetic recording began in 1898, when Valdemar Poulsen, a Danish engineer, gave a public exhibition of his Telegraphone. This instrument consisted of a piano wire 1.5 metres (59 inches) in length, along which was passed an electromagnet in close contact. The coil of the magnet was connected in series with a battery and carbon microphone from a telephone. The piano wire became magnetized by speaking into the microphone, and by substituting a telephone receiver the speech could be replayed. But it was 46 years later that Germany produced a tape recorder, the Magnetophon, to relay Hitler's messages to the army. After further development of the system in America, magnetic sound was introduced into the film industry during 1950. New techniques had to be learned for handling and synchronizing magnetic film, since recordings were invisible and unlike optical soundtracks to which everybody had become accustomed. The improvements obtained were virtually silent backgrounds, extended frequency response, immediate replay after recording, and quick methods of making copies.

All soundtracks nowadays originate on ¼ inch magnetic tape using portable recorders such as the Nagra, Perfectone, and Stellavox. In addition to the audio track, a separate pulse track is also recorded, which provides a speed reference of the picture camera. The selected material is transferred on to magnetic film for editing purposes, and the pulse track is used for synchronizing the tape to the magnetic film recorder. A wide magnetic stripe on clear 35 mm film base is preferred for editing, but fully coated magnetic film is used for recording up to 6 tracks on 35 mm film.

The component parts of a soundtrack, meaning dialogue, music, and effects, are synchronized to the edited picture and assembled into separate reels of magnetic film. There may be as many as 15 or 20 reels of sound for one reel of picture, and these are mixed together in a re-recording theatre equipped with a long sound mixing console. Two or three balance engineers adjust volume levels and sound quality to the satisfaction of the producer, and a final three track magnetic master is recorded with dialogue on track 1, music on track 2, and effects on track 3. This allows for other languages to be substituted and mixed with the music and effects for different countries. The magnetic master is then re-recorded on to an optical sound negative for making standard release prints.

Some release prints have magnetic soundtracks, in which case a magnetic coating consisting of an iron oxide lacquer is applied to the film base after printing. 35 mm films have three magnetic stripes, or four stripes when small perforations are used. 70 mm films have six stripes, five of which feed loudspeakers behind the screen, and one feeds auditorium loudspeakers for 'surround sounds' and special effects (such as low-frequency earthquake sounds).

16 mm films may have perforations down both edges or just down one edge. In the former case, edge stripes have to be used—one thin stripe down each edge. In the latter, one edge of the film can be used for the soundtrack, which may be either all magnetic, all optical, or half and half—offering the possibility of dual language soundtracks. These are frequently used on airline services offering in-flight movies. In all cases where stripe is added to the film after editing, a thin balance stripe must be added along the other edge so that the film will be the same thickness at both edges and will not wind up awkwardly on the spool.

Projectors for 35 mm and 70 mm films have penthouse sound heads—that is, they are between the upper spool box and the picture gate, rather than below the gate. This gives a picture to sound separation of minus 24 frames for 70 mm and minus 28 frames for 35 mm. With 16 mm films the sound reproducer is below the picture gate, and the sync separation is 28 frames.

OTHER SYSTEMS New developments in soundtracks include a hue-modulated sound negative recorded on colour film. The system has signal peaks of red and green, and yellow for zero modulation. Experiments with stereo optical are being made with a twin-track Dolby encoded sound negative. A decoder is used for replay as three track stereo. Quintophonic consists of a three track magnetic stripe on 35 mm film with quadraphonic encoding. This can be replayed as five separate tracks. The Imax system uses 70 mm picture film projected horizontally, with separate six track magnetic sound replayed through 55 loudspeakers.

AMATEUR SOUNDTRACKS The first amateur sound films were recorded on 78 rpm discs coated with cellulose lacquer. These discs had the advantage of immediate replay, but were rather fragile and had a short life. Magnetic recording gave improved and permanent sound quality, and numerous mechanical and electrical devices were designed to synchronize a tape recorder to a projector. The various systems evolved meant that films were not interchangeable, and synchronization was not always good.

The preferred method today is a magnetic stripe in liquid form, which is applied to the edited film. The soundtrack is recorded later on a projector, or sometimes during filming within the camera, with magnetic facilities, which gives good sound quality with a frequency range of 50 Hz to 8 kHz. Automatic volume control is a common feature, and sometimes automatic mixing of music and commentary is available. All 8 mm magnetic projectors record at a standard picture to sound sync separation of 56 frames on standard 8 mm, and 18 frames on Super 8 mm. These systems usually allow only approximate synchronization, such as for mood music or commentary, and full lip-sync—that is, voices linked to the lip movements on the screen—is not possible.

Many amateurs increase the complexity of their soundtracks by following professional practice, shooting lip-sync dialogue scenes on reel-to-reel or cassette recorders, and recording a reference pulse track from the cine camera drive mechanism. Numerous devices are available for transferring the sound on to the striped picture film, and some equipment also allows for a limited amount of sound and picture editing, resulting in films which approach professional quality.

Above: powdered iron oxide, the same chemical formula as rust, is an important pigment used in paint manufacture (it is also used for the cosmetic rouge). Other metal compounds used in the paint industry include those of lead, zinc, cadmium, copper and titanium, each giving a characteristic colour, as well as many other compounds.

PAINT manufacture

The beginnings of the making and use of paint by man go back far into prehistory. The activity probably arose from his desire to decorate the walls of his home with reminders of things he prized. Animals, important as a source of food and skins, have been found depicted in cave dwellings at Lascaux in France and Altamira in Spain.

These paintings were probably produced by smearing the rock with a finger dipped in a rudimentary mixture of coloured earth and water or perhaps animal fat. These simple constituents—coloured powder and fluid carrier—have remained through the ages essential ingredients of paints.

Paint can be defined as any fluid material that will spread over a solid surface and dry or harden to an adherent and coherent coloured obscuring film. It usually consists of a powdered solid (the *pigment*) suspended in a liquid (the *vehicle*, *medium* or *binder*). The pigment provides the colouring and obscuring properties. The binder is the film-forming component which holds the pigment particles together and attaches them to the surface over which they are spread.

Decoration seems to have been the original purpose of paint, but in time its power to protect the vulnerable surfaces of manmade objects became of almost equal importance.

PAINT SYSTEMS AND TYPES For maximum protection and durability it is necessary to employ a

Above: mixing the powdered pigment into the base. Thorough powdering of the pigment and complete, even mixing of all ingredients are of the highest importance in the manufacture of any type of paint.

Below: one of the main stages of paint manufacture is the dispersal of pigment within the base liquid. This is commonly done in a roll mill like this one, the paint being removed from the front roller with a scraper blade.

multiple coat system. The first layer of primer ensures adhesion between the substrate and subsequent coats. Then comes the undercoat to obscure what is beneath it and provide a suitable surface for the final coat. Finally there is the topcoat or finish to give the required appearance as regards colour and gloss. Paints are available to give a full gloss, 'eggshell' or matt finish.

The traditional oil gloss paint was a mixture of linseed oil, pigment, thinning solvent (commonly turpentine) and additives to promote drying. Addition of natural resin to refined and treated linseed oil improved the spreading properties, rate of drying and gloss. The blend of resin and oil became known as varnish and further developments led to longer-lasting 'enamel' or 'hard gloss' paints. Another type that has become increasingly popular in recent years is emulsion paints, in which the vehicle is a suspension or dispersion of a material such as polyvinyl acetate in water.

A widely used alternative to linseed oil, when the price is favourable, is soya bean oil. The resulting films show a much reduced tendency to yellow on exposure to air and moisture, and are therefore favoured for white and pastel colours. Other non-yellowing oils sometimes used are those extracted from the seeds of plants such as the tobacco, the safflower, the sunflower, and the poppy—a traditional favourite for artists' oil colours.

RESINS AND PIGMENTS The use of natural resins such as Congo copal in coatings is of great antiquity but of little importance today. For the past 30 years or so the paint industry has widely used synthetic oil-modified alkyd resins which are essentially derived from the reaction of glycerol with phthalic anhydride. Alkyd resins may also be modified with other synthetic monomers (single molecules) to give, for example, the alkyd-melamine finishes used for several years by parts of the British motor industry.

Recent years have seen increasing use by the paint industry of other types of synthetic resins. Acrylic finishes based on polymers such as polymethyl methacrylate are used in several countries in durable building paints and vehicle finishes. Epoxy resins, typically derived from bisphenol and epichlorhydrin, are probably best known for their use in adhesives but are also used in corrosion-resistant coatings. Another category of growing importance is the polyurethane resins based on tolylene di-isocyanate and used in tough coatings. The continuing use of one natural resin, nitrocellulose (in a low nitrogen content form), in furniture lacquers and related finishes should not be overlooked.

For many years the principal white pigments were white lead (basic lead carbonate), zinc oxide and lithopone (a mixture of zinc sulphide and barium sulphate). These have now been almost entirely superseded on grounds of opacity, hiding power and toxicity by titanium dioxide, used in one of its three crystalline forms known as rutile. This is because rutile is free from the defect of the anatase crystalline form which causes paints to 'chalk' (lose gloss by developing a loose powdery surface). The paint industry consumes about 65% of the growing world output of the pigment.

In corrosion-resistant primers for iron and steel, red lead continues to be of prime importance. Lead chromes are widely used for yellow and orange shades, and in combination with blue pigments as the basis for greens. Cadmium sulphide has long been used as an artists' yellow, while metallic aluminium, zinc and lead have specialized uses.

All the pigments mentioned so far are of inorganic or mineral origin. Largely as a result of demand from the textile industry, a wide range of organic pigments is now available and finding increasing application in paints. Of particular interest are the metal-complex pigments such as the blue copper phthalocyanine. Another organic pigment is carbon black, derived from vegetable or mineral sources.

PAINTMAKING Manufacture of paints—the dispersion of a pigment in a vehicle—requires two kinds of movement. These are squeezing and rubbing, or in technical language pressure and shear. Early artists mixed their paints between glass or granite surfaces in a simple muller operated by hand. The first mechanized milling devices were two flat stones one on top of the other, the top one rotating and the bottom one fixed. An improvement on this flat stone mill was the cone mill and other types, some of which require the charge to be pre-mixed.

A widely used unit is the triple roll mill. In a typical installation the rolls are of 12 inch (30.5 cm) diameter and 30 inch (76 cm) long. They rotate at different speeds, the front rolls being fastest. A typical speed ratio is 1:3:9, the back roll moving at about 30 revs/min. The speed differences provide the rubbing and squeezing actions and also give a more evenly spread film. Extra shearing effect is produced in some mills by a sideways sliding motion. A scraper blade removes the paste from the front roller, when it may be ready for packing or for further thinning.

Many paintmakers use various sizes of ball mills, which can produce well-finished gloss paints with a minimum of supervision. The machine consists of a rotating cylinder which contains steel or porcelain balls or flint pebbles in quantity sufficient to occupy about 45% of the total volume. The actual space occupied is about half the apparent space and it is in the interstices that the paint is milled. The optimum point charge is therefore about 20% of the total volume, while a typical milling cycle occupies about 16 hours (that is, overnight).

A logical development from the ball mill is the modern sand mill, in which the speed of rotation is no longer limited by the centrifugal action of the grinding media. A sand mill consists essentially of a vertical cylinder containing sand of very small particle size

Above: hand grinding a paint sample in the laboratory to check that the colour balance is correct before large-scale production begins.

Left: paint samples being exposed to ultra-violet light for two weeks to test colour fastness. This test is the equivalent of six month's sunlight.

Right: the 'make-up' stage of paint manufacture.

or fine glass beads (ballotini), driven at high speed by a multi-disc impeller. A pump forces the pre-mixed paste of pigment and vehicle through a non-return valve into the milling area, where dispersion takes place by the rubbing of the paste between the fine particles of sand. The product is discharged through a fine wire mesh to retain the sand.

A notable feature of the sand mill is that it allows continuous operation. Paint manufacture has traditionally been a batch process, but large modern plants now operate on at least a semi-continuous basis for the most popular finishes and colours.

Another modern development is pigments which can be dispersed by mixers with specially designed blades rotating at very high speeds. In these high speed impeller mills, dispersion is again effected by attrition of the pigment aggregates and the consistency achieved may be fairly high.

HEALTH AND FIRE HAZARDS Many of the materials used in paint manufacture are potentially hazardous if mishandled, but the risks are generally well understood and only elementary precautions are required of the user. Among pigments, the compounds of lead and other heavy metals have attracted particular notice because of the hazard to young children with the habit of chewing or sucking flakes of old paint. As noted above, lead pigments are now little used in decorative paints, but some countries have organized action to remove or cover up old lead-based coatings. It should be understood, however, that a small proportion of lead compound is necessary for the rapid drying of modern paints and could only be completely eliminated at substantial cost in both price and performance. Appreciable levels of lead continue to be used in industrial finishes, where application by spraying is subject to control.

The solvent benzene is now rarely used because of danger to the lungs. Precautions are necessary with chlorinated solvents, and adequate ventilation is recommended with all spraying operations. It is also often advisable to avoid contact with the skin or eyes.

Certain specialized materials present unusual hazards. Thus nitrocellulose (another form of which is also known as gun-cotton) is classified as an explosive, while the recently introduced powder coatings ('dry paints') are potentially capable of giving rise to dust explosions.

Most organic substances are combustible, that is, they will inflame if heated strongly enough. The flashpoint, which indicates the degree of flammability of a solvent, is the lowest temperature at which the vapour collected in a closed space ignites on the introduction of a small flame. The precautions to be observed depend on the flashpoint and may involve the use of flameproof electrical equipment. Particular care may also be necessary in storage and waste disposal.

POLLUTION CONTROL In recent years industry as a whole has become increasingly conscious of its

responsibility to avoid damage to the environment. This has led, in the paint industry, to close control of solid and liquid wastes. There has been some interest in the recycling of solvents, or their elimination in solventless coatings or powder coatings. There has also been a marked trend to water-based coatings and much effort has been devoted to the search for a full gloss emulsion paint.

A side effect of pollution control has been some scarcity of titanium dioxide. The widely used sulphate process for its manufacture gives rise to large quantities of coloured effluent and some countries have restricted output until steps are taken to avoid damage to the environment. Carbon black is another important material subject to control on ecological grounds.

RECENT DEVELOPMENTS For many years the most popular methods for applying paints were brushing, dipping and spraying. The last decade has seen widespread adoption by the motor and other industries of the electrodeposition or electropainting process, which involves passing an electric current between

Right: filling paint cans. The paint is fed into the machine through pneumatically controlled valves, and dispensed into the cans, which are then fitted with lids and packed ready for dispatch.

Below: spraying a car body with acrylic enamel paint. Before the enamel is applied the body is coated with primer by a process in which the car is negatively charged and the primer is positively charged, the paint being deposited electrostatically.

the metal article to be coated and the tank in which it is immersed.

Drying methods have progressed from natural evaporation of solvents, through forced drying in heated ovens to radiation curing, whereby specially formulated coatings 'set' in a fraction of a second under the influence of infra-red or ultra-violet radiation.

Another modern development is coil coating, in which acrylic, epoxy and other finishes are roller coated on to continuous steel strip before fabrication into building cladding or domestic appliances.

The modern paint industry is firmly based on scientific principles and employs all the techniques known to the chemical industry as a whole. Computers are employed in the complicated area of colour measurement and colour matching, notably on the numerous and frequently changing colours used by car manufacturers. International standardization of test methods, for example on accelerated weathering, is well advanced, while the European Economic Community has drafted a directive on the labelling of paints.

Above: papermaking in Thailand in the traditional way. The screen has just been dipped into the solution.

Below: trimmed, peeled logs in the grinding machine. Gravity causes the logs to fall on to grinding wheels which are flushed with water. Most paper in Britain is made from wood.

PAPERMAKING

Papyrus, a water reed, was used as a writing material more than 5000 years ago by the Egyptians, and later by Greek and Roman writers. Animal parchment has been known since Egyptian times and was much favoured by the Romans for permanent records. To make papyrus, layers of reeds are set across each other and pressed and dried; parchment manufacture is similar to leather manufacture, involving the scraping and treatment of animal skins (except that nowadays certain grades of paper are called vegetable parchment). Neither of the ancient processes changed the structure of the basic materials.

The word 'paper' comes from the word papyrus, but true paper was invented in China less than 2000 years ago. The Chinese collected old fishing nets, rags and bits of plants, boiled the materials well, and beat them and stirred them with large amounts of water to make a pulp. A sieve dipped in the pulp and removed horizontally would have a layer of pulp on it with the water draining away through the mesh. The layer of pulp was then dried and pressed. The difference between papyrus and paper is that in papermaking the materials are reduced to their fibre structure, and the fibres are re-aligned.

The technique spread to the West when some Chinese papermakers were captured by the Arabs. It reached Europe in the late mediaeval period, and the first English paper factory was established in Hertfordshire in 1490. Paper made possible more literacy, and as literacy spread there was increased demand for paper. In the 20th century the latest advances in papermaking have been exported from America to Japan, so that the technology has made a complete circuit of the globe.

MATERIALS Rags are used for the highest quality paper, especially the handmade. At the other end of the scale, seed fibres, jute, flax, grasses and other plants may be the source of raw material for papermaking; straw was used extensively in Great Britain during World War II. The largest amount of paper today, however, is made from wood pulp.

Synthetic and animal fibres have also been tried, but these techniques have remained experimental or too expensive for wide application; papermaking means almost exclusively the use of cellulose vegetable fibres. The fibres vary in size and shape, but are hollow tubes closed at the ends and often tapering. They are held together in their natural state by substances, principally lignin, which must usually be dissolved and removed; this is accomplished by chemical treatment and washing of the pulp. The wide variety of methods found in papermaking can be seen in a comparison of blotting and greaseproof wrapping papers: the one is

Left: the ground wood being transferred to the digester.

Below: schematic diagram of the process for making paper from wood. Logs are peeled and mechanically pulped or cooked with chemicals. Waste paper is also widely used. The mixture is further treated mechanically so that the fibres are the right length; the necessary additives are mixed in, and the pulp is poured on to a moving wire belt which is shaken to align the fibres. The sheet of pulp is pressed and dried by a series of rollers, and the finished paper is then reeled up.

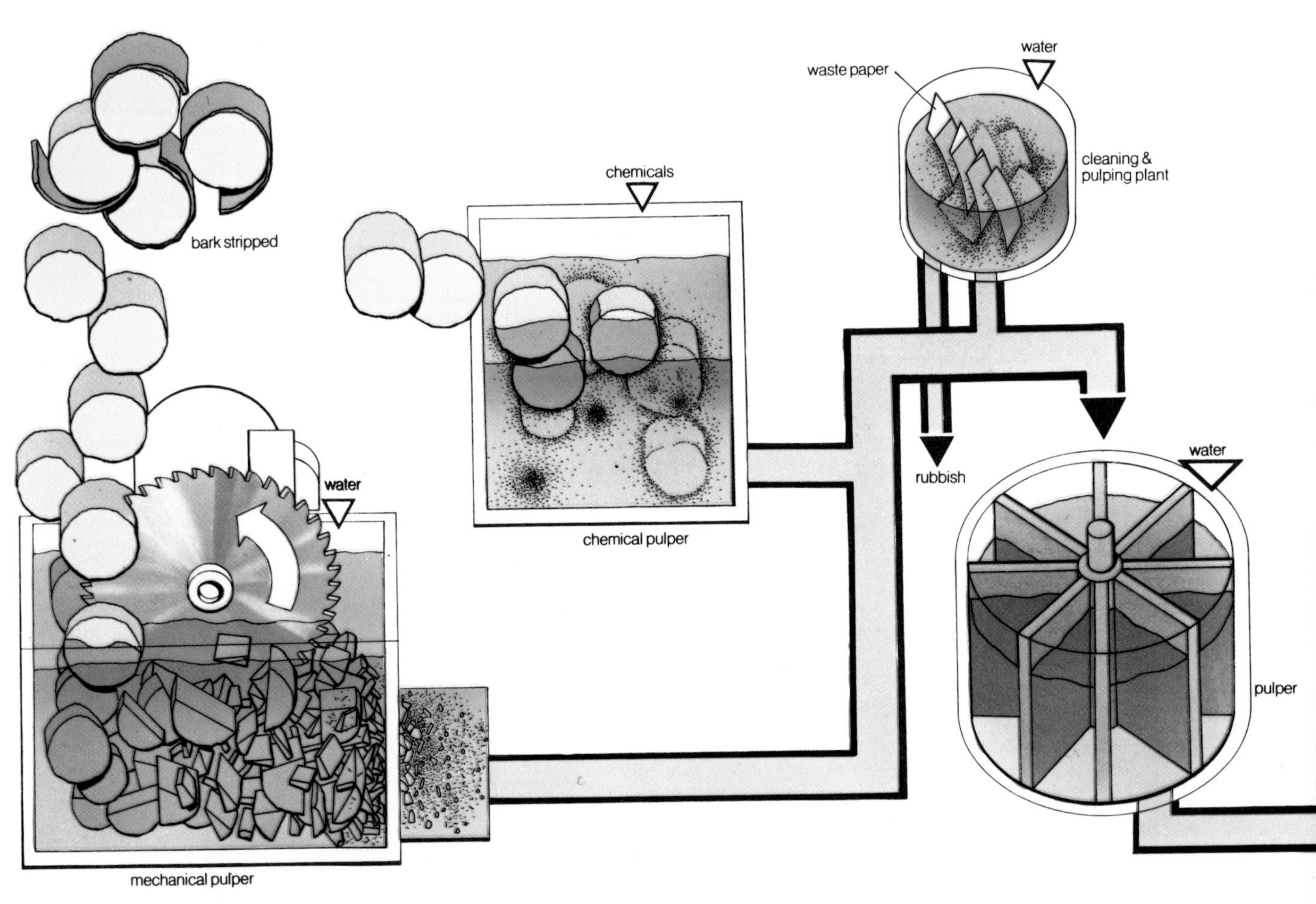

soft and absorbent, while the other is hard, smooth and dense. The difference is in the choice of fibres, the way they are prepared, and the way they are processed on the papermaking machine.

PULP MANUFACTURE There are basically two methods: mechanical and chemical (cooking).

Mechanical pulping is used chiefly for coniferous woods. It aims at a high yield rather than a pure pulp; the result is a cheap paper, of which newsprint is a good example, which is not expected to last. The logs are trimmed, de-barked and then ground, usually by rotating grindstones, the fibres being flushed away from the stone with water. If the water supply is lowered, more heat is generated and longer fibres are obtained; in general, shorter fibres (up to a point) result in better paper. The pulp is screened several times, and the larger lumps are re-treated or burned. Depending on whether the pulp is to be processed on the spot or shipped, the excess water is removed in a concentrator or on a machine resembling a simple papermaking machine (see below). The result is either air-dry (10% moisture) or wet or moist (45% moisture). 100 tons of dry cut logs can yield more than 90 tons of air-dry pulp, but the strength of the fibres is not high and mechanical pulp is mixed with 15% to 50% chemical pulp before it is used.

Chemical or cooking methods remove more of the unwanted materials, resulting in a lower yield but a higher quality pulp. They are divided generally into two categories: acid liquor and alkaline liquor.

The acid liquor process is used mostly for spruce, which is the largest commercially profitable tree crop in North America. The liquor is essentially an acid bisulphite with some free sulphur dioxide gas. It can be made by letting water trickle down through a tower containing limestone and blowing in sulphur dioxide gas at the bottom. It is highly corrosive, which means that the works must be made of acid-resistant materials. The logs are sliced and the slices broken up into chips which are then screened. Chips from $\frac{1}{4}$ to $\frac{3}{4}$ inch (0.6 to 1.9 cm) are pressure-cooked in a steam-heated digester. The quality is controlled by regulating chip size, liquor-strength, pressure (520 to 760 kN/m^2, 75 to 110 psi) and cooking time (usually 7 to 12 hours). The unused sulphur dioxide and some of the heat can be recovered when cooking is over, but the spent liquor is highly polluting. It can be used in leather manufacture as a tanning agent, to lay the dust on roads and in several other ways, but so much of it is produced that it is still a serious pollution problem.

Alkaline liquor processes are similar but the cooking agents are naturally not acidic and therefore less polluting. They are for non-woody fibres such as grasses and rags, for deciduous woods (hardwoods) and for coniferous woods which have a high resin content, such as pine, because the alkali dissolves the resin. The wood is prepared in the same manner as for acidic cooking; other materials are prepared ac-

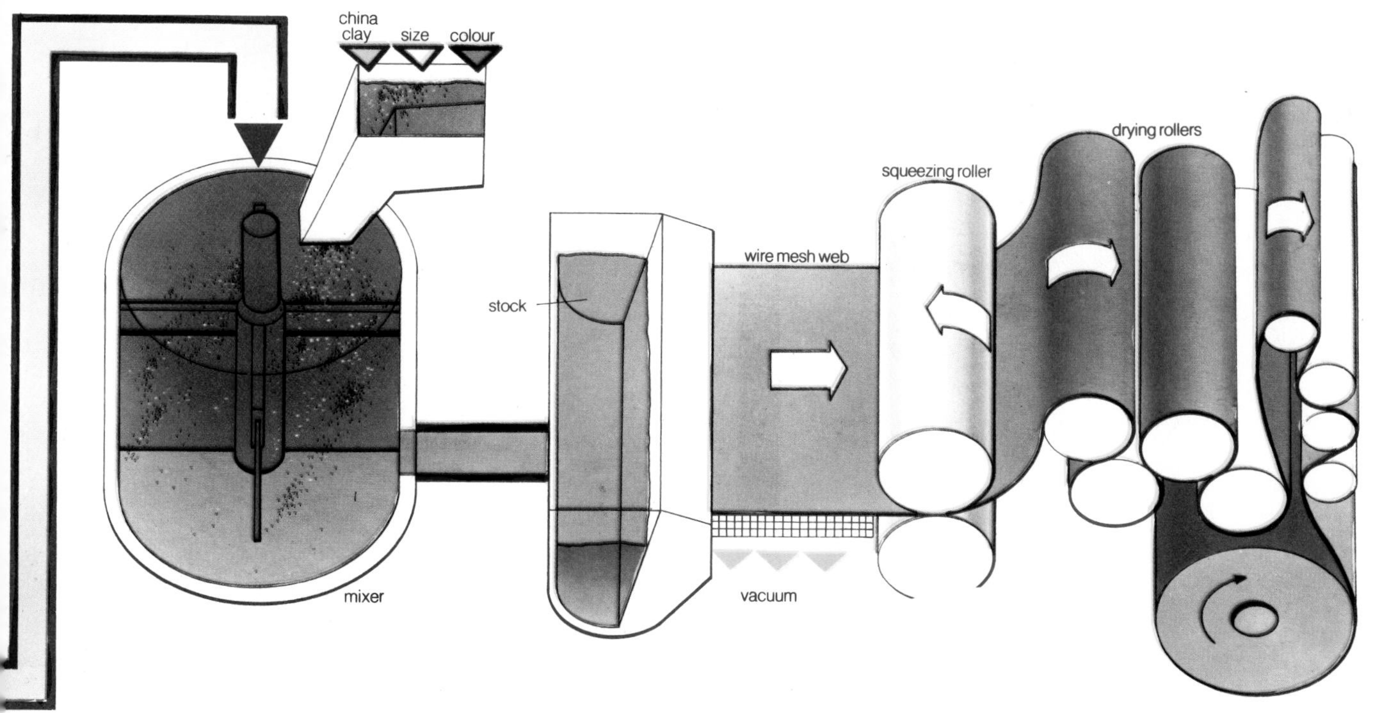

Below: wood pulp being prepared for the papermaking machine; this controls the length of the fibres.

Bottom: the dry end of the continuous process papermaking machine. The paper has just gone through the drying train and the calender rolls; the moisture and static electricity content have also been adjusted. It may now be slit or cross-cut into sheets.

Opposite page: the wet end of the machine. The moving belt is shaken to align the fibres; a vacuum below it sucks out the water. Then it goes around rollers on a belt of felt and more water is squeezed out.

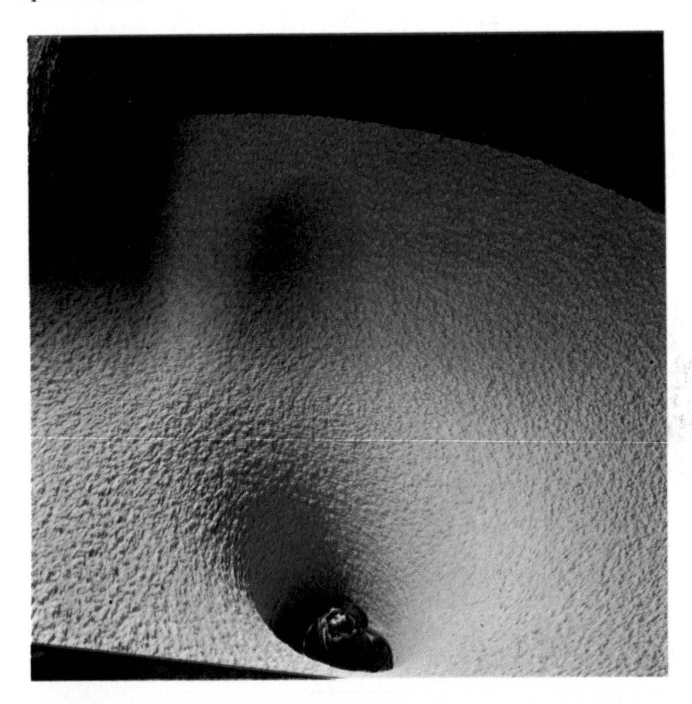

cording to their properties. For example, rags must be sorted and straw must be chopped and the dust removed in a cyclone extractor.

The soda process uses caustic soda (sodium hydroxide). The amount of soda, cooking time and pressure all vary according to the materials being cooked. With wood, up to 85% of the soda can be recovered from the waste liquor; with other materials less is recovered because of the difficulty of washing it out of the fibres and because the amount of soda used is smaller to begin with. The kraft or sulphate process results in a stronger fibre; 'Kraft' is the German word for strength. Sodium sulphate is added to the digester; it has no effect on digestion, but is converted to sodium sulphide during the burning of the recovery process. When this in turn is added to the digester it is automatically converted in controlled amounts to sodium hydroxide, so that it aids in digestion but conserves strength. The process generates objectionable smells and is not used near towns; it results in a scum on the waste liquor, which is called tall oil, and can be used to make soap and lubricants. Sodium monosulphite, also called the neutral sulphite process, is one of the newer cooking methods. It depends on sulphur dioxide and caustic soda and results in a high yield, and can be used to treat hardwoods which were not previously suitable for papermaking. The recovery process is still under development.

Continuous digesting is also under development, and there are combined mechanical and chemical methods in use which do not need pressure but lend themselves to continuous operation. For example, straw can be treated in a suitable vessel with an impeller and a solution of hot soda, resulting in a high yield of straw pulp for making packaging grades.

PULP PREPARATION After digestion the pulp must be washed, and often bleached. Washing is necessary to get rid of impurities; some pulps can be washed while still in the digester, and this is part of the waste recovery. The first wash results in a strong liquor which can in some cases be recovered; the next wash will result in a weaker liquor which can be used for the first wash of the next batch, and so forth.

With wood pulps and in high capacity operations the digester is emptied immediately so it can be used again. Screens in series, both flat metal tray types and rotating devices, are used to remove impurities, but centrifugal or vortex cleaners are now so efficient that they can replace screens altogether. In a vortex cleaner, as the pulp is rapidly rotated heavier impurities fall to the bottom while pure pulp passes through outlets near the top. One widely used centrifugal machine is called a rotary vacuum filter; this comprises a wire drum revolving in a vat. Suction from inside the drum draws a layer of pulp on to the wire, sucking the liquor out. The pulp is then washed with hot water sprays and scraped off the wire by a doctor blade.

Next follows, if required, some type of bleaching process, depending on the type of pulp and its intended use. Methods vary from the simple addition of bleaching liquor to a series of chemical steps which must be carefully monitored to avoid damaging the fibres. Increasingly nowadays the bleaching is an extension of the digestion process, the exact process being a combination of methods chosen to accomplish maximum yield, strength of fibres and whiteness.

BEATING In most cases the pulp arrives at the paper mill in sheet form and must be broken up again. For this purpose a machine called a hydrapulper may be used, a tun-shaped vessel with an impeller which disintegrates the pulp in water. Similar treatment is applied to waste paper, an increasingly important raw material, particularly in view of recent increases in the price of pulp.

Next comes the beating. Prolonged beating with the traditional machinery is still practiced, especially for the handmade papers, but highly developed quality control followed by treatment in machines called refiners increasingly fulfils the beating function. In any case, the quality of the finished product is determined more at this point than at any other. The physical action of beating affects the length of the fibres, their plasticity and their capacity for bonding together in the paper machine; therefore beating also determines such characteristics in the paper as bulk, opacity and strength.

The most common type of beating machine is the hollander, developed in Holland in the 18th century. It is an oval-shaped tube which has a low wall called a mid-feather running across the centre, but stopping short on each side so that the pulp can circulate around it. On one side of the mid-feather the beater roll is mounted on a shaft; the beater roll may weigh as much as ten tons and the capacity of the tub may vary from about 90 kg (200 lb) of rags to about 1½ tons of wood pulp, with perhaps five parts wood to 95 parts water. The roll has bars on its circumference parallel with the shaft; the pulp, now called the stock or the stuff, is ground against stationary bars on the floor of the tub beneath the roll. The clearance between the roll and the stationary bars is small but adjustable.

The refiner, which has completely superseded the beater in the making of newsprint, for example, is a cone-shaped beater roll in a similarly shaped housing, also equipped with bars and adjustable clearance; the bars in the refiner run at a speed of 3000 feet (over 900 m) per minute. The stock goes through the refiner only once, and the refiners are connected in series if further beating capacity is needed. There are also disc refiners, with one rotating disc and one stationary, and combination cone and disc models.

Loading or fillers, pigments or dyes if required, and most sizing agents may be added to the stuff during beating. Loading materials are added to give improved opacity, and they also help to make the paper stable dimensionally and assist in obtaining a good finish. They are white materials, of which the most common are china clay, titanium oxide and precipitated chalk. Chalk, for example, is added to cigarette paper to make it burn more evenly. Sizing agents, of which resin is the most common, render the

paper resistant to penetration by water (but do not make it waterproof) so that it can be written upon with water-based writing ink. Printing inks are oil or spirit based, and so papers for printing do not need to be sized, but completely unsized paper, called waterleaf, is not common. Wrapping papers may have to be written upon, for example, and certainly papers used in lithography should have some water resistance.

PAPERMAKING MACHINES The actual papermaking process is a continuous one. The most common type of machine is the Fourdrinier, named after the two brothers, stationers, who built the first one in Hertfordshire in 1803. Their machine deposited the paper on to pieces of felt, after which it was finished by hand; the modern machine starts with the dilute stock at the wet end and finishes with reels of paper at the dry end. The additives (loadings and so forth) can be mixed into the stock in the wet end of the machine, instead of in the beating, if desired.

The stock is continuously delivered on to an endless belt called the cloth which is made of a wire or plastic mesh. A short sideways shake is applied to the cloth where the stock first meets it, to improve the way the fibres mesh together. Drainage of water begins immediately through the mesh of the cloth, bringing the fibres closer together until the stock becomes a cohesive web. Then suction is introduced by means of vacuum boxes underneath the web; at this point also, a light wire-covered roll called the dandy roll rides on the upper surface of the web, usually turned by the travelling web (but sometimes turning under power) and gently pressing it. A wire design can be wired or soldered to the dandy roll to impress the watermark of the manufacturer on to the web.

The web is now separated from the cloth by a pair of rollers, or, on the latest model machines, by a single suction roller. These are called couch rolls. The volume of water removed up to this point is rich in fibre, chemicals and so forth, much of which is recovered. The web is deposited on a felt and carried between pairs of pressure rolls which remove more of the water; the felt frequently becomes clogged and must be cleaned.

The water content of the web has now been reduced to about 65%, and the web passes to the dry end of the machine, where it passes around a series of pairs of steam-heated iron cylinders. At the end of the drying train, the paper is without a finish and has been overdried to about 3 to 4% moisture content. The paper passes through several calender stacks, which are pairs of highly polished chilled iron pressure rolls. The stacks smooth the surface, sometimes using a little water to do so. Finally the paper passes over cooled sweat rolls, which adjust the moisture content and reduce the static electricity, and is wound on to large reels. To avoid problems with dimensional stability, many printing papers are conditioned on the reel by passage through a hot wet atmosphere which adjusts the moisture content to 6 or 8%, which is normal for this type of paper when mature.

The paper is slit to the width and reel diameter required, if it is to be used from the reel, and paper to be used in sheet form is taken to a cutting machine where it is slit and cross-cut in the same operation. For greater accuracy and clean edges, some papers are afterwards guillotine trimmed. Sheet papers are generally inspected, torn or faulty sheets removed, and counted for packing; nowadays this can be done electronically at the cutting stage.

OTHER MACHINES The MG machine (for Machine Glazed) produces paper which is highly glazed on one side and rough on the other. The paper is used for posters, general wrapping and carrier bags; the glaze is imparted by a large, highly polished drying cylinder. The paper is stuck to the face of the cylinder by a pressure roll and the surface in contact takes on the polish of the metal.

Boards (cardboard) for cartons, packaging and so forth are made by a different method of sheet formation on a cylinder mould machine. A wire covered roll rotates in a vat of dilute stock. Water filters through to the inside of the roll and a layer of fibres is left on the surface which is transferred to a felt. The process makes only a thin layer; to make up the greater thickness of board, several such moulds are placed in series so that a multi-layer structure is built up on the underside of a making felt, which is then reversed and carried through the rest of the machine. The greater advantage of this type of machine is that the centre layers of board can be made entirely of cheaper materials, such as waste paper.

Nowadays suction devices are becoming more common to assist the couch and other rollers on papermaking machinery. Some machines have the dry end entirely covered by a hood, within which fans and pipes remove moisture-laden air, enabling the machine to run faster. A modern fast newsprint machine will run at around 1000 m/min (3280 feet per minute).

PHOTOGRAPHIC processing

Exposing photographic materials normally does not produce a visible image, merely an invisible or latent image consisting of silver bromide crystals, some of which contain specks of silver produced by the action of light. Processing is the name given to the sequence of chemical treatments needed to convert the latent image into a stable, visible one. There are only a few basic different steps: development, fixation (or stabilization) and bleaching, but a wide range of different effects can be obtained by combining the steps in a different order or carrying them out in different ways.

The first, and most important, step is development, in which exposed silver halide grains are reduced to silver while unexposed grains are not affected. The ability of a small latent image centre, which is formed during exposure and may consist of only about 10 atoms of silver, to initiate the reduction of the whole grain, which may contain up to about 10^{10} (10,000,000,000) atoms of silver, is the fundamental reason for the high sensitivity of the photographic system.

A developer therefore contains a suitable reducing agent or mixture of two agents. Hydroquinone, *p*-aminophenol and *p*-phenylene diamine are typical members of three different classes of developing agents commonly used. Since the reducing power of such agents increases with the alkalinity of the solution, it is controlled by incorporating a suitable alkali, such as borax, sodium carbonate or caustic soda, in the developer. Since the bromide ions released by the development reaction reduce the rate of reaction, potassium bromide is usually incorporated to minimize the difference between fresh and used developer. It also decreases the tendency to reduce some unexposed grains.

Developers also contain sodium sulphite which acts as an anti-oxidant, suppressing the oxidation of the developing agent by atmospheric oxygen and thereby extending the shelf-life of the developer. Sulphite also reacts with the oxidized developer, which is unstable in alkaline solutions and decomposes to give coloured products, to give colourless products which do not stain the photograph.

In black and white photography, the silver, which is formed as a tangled mass of very fine filaments and appears black, is the primary product and the oxidized developer is removed during subsequent processing steps. In colour photography, on the other hand, the silver is unwanted while the oxidized *p*-phenylene-diamine developer produced in the development reaction reacts with colour coupler to form an image dye. In normal development (both black and white and colour) no reaction occurs in unexposed areas and most darkening is produced in the most fully exposed areas. The image therefore is a negative; one in which the tonal values are reversed.

FIXING The silver halide that is not reduced during development is slightly coloured and, being photosensitive, darkens further on prolonged exposure. It must therefore either be removed from the layer (fixation) or converted to a colourless inactive material (stabilization). Fixing agents are compounds which form soluble complex salts with silver compounds and which therefore dissolve silver halides as a complex salt. Fairly concentrated solutions of sodium or ammonium thiosulphate are usually used; the former compound is frequently called 'hypo', an abbreviation of its old name sodium hyposulphite.

Fixing solutions used in black and white processing are usually made acid to diminish the reducing power of any developing agent carried over from the developing tank which might otherwise reduce part of the silver halide as it dissolves to produce dichroic fog. Such baths also contain sulphite to diminish the decomposition of the thiosulphate into sulphite and sulphur. After fixation, the film or paper must be thoroughly washed as any residual fixer left in the print decomposes to give coloured products and attacks the developed silver on long storage.

Stabilization is intrinsically less satisfactory than fixation. The final prints are less stable and, since they contain insoluble white silver compounds, are confined to prints on paper. The process is used when convenience and speed of access is more important than the quality of the final image, and is commonly used in print processing machines. Stabilizers are usually organic sulphur compounds. After development, the

Below left: roll films are usually developed in tanks either individually or in frames such as this. Each film is held by the edge in a stainless steel spiral. The frame is transferred from tank to tank in the dark, so this picture does not show a real-life operation.

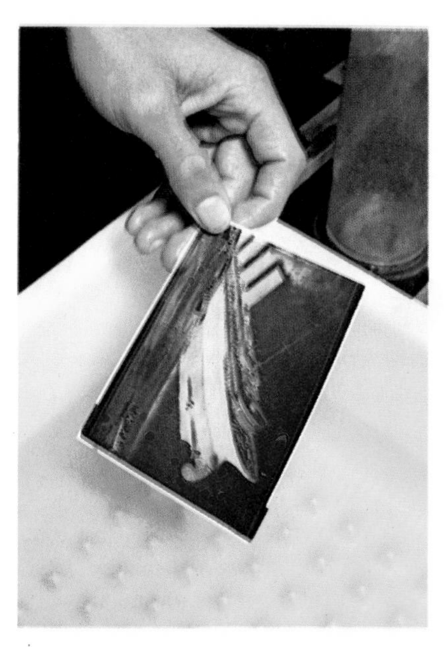

Below centre and right: the next four pictures show stages in film processing. A copy negative has been made on sheet film and is being developed in a dish. The first picture shows the film's appearance after a short while in the developer, then when fully developed.

Right: continuing the sheet film process – the developed negative still has opaque white emulsion, so that the image is hardly visible from the back. The first picture shows it half in the fixer, so that only the lower part has cleared. After fixing, it is washed and hung up to dry, as shown in the last picture.

unreacted silver halide is converted into a white inactive compound by bathing in a solution of the stabilizer. Stabilized prints are not washed as this would remove the excess stabilizer and make the print less stable. A stabilized print will, however, last for several years without discoloration; if at any time a permanent print is wanted, it can be fixed and washed in the normal way.

BLEACHES Bleaches are used to eliminate developed silver from photographic materials. Although very many different formulations are used, they fall into three main classes. Simple bleaches, such as a solution of potassium dichromate and sulphuric acid, convert the silver to a relatively soluble silver salt (silver sulphate) without affecting the undeveloped silver halide. Rehalogenizing bleaches contain a milder oxidizing agent, such as potassium ferricyanide, together with potassium bromide and reconvert developed silver back to silver bromide. They are frequently used in colour processing, where the reformed silver bromide is removed in the fixing bath which removes the undeveloped silver halide.

Finally, bleach-fixing solutions contain an oxidizing agent, such as thiosulphate. They remove both developed silver and undeveloped silver halide simultaneously and are used to eliminate a step in some colour processes.

NEGATIVES AND REVERSAL Photographic systems in which the original photograph is used to give an enlarged print, such as most black and white photography, or in which one master is used to make several copies, such as most professional cinematography, necessarily involve two photographic steps which may either both involve negative processing, with light and dark tones interchanged, or may both involve reversal processing in which the film is processed to give a positive result, with the tones correct. Negative processing is usually used because it is simpler. For black and white photography, development is followed by fixing, washing and drying, and for colour, colour development is followed by a bleach step to remove the developed silver before fixing and washing.

Reversal processing is usually reserved for photographic systems in which the original photograph is the final product, such as in colour transparencies for projection and amateur cinematography. In these cases the increased complications in processing are compensated for by the elimination of a second photographic step. In black and white reversal processing, the exposed emulsion is first developed to give a negative image and then bleached in a simple bleach. The undeveloped emulsion grains are fogged, or rendered developable, by exposure or by chemical means and then developed in a second developer. In colour reversal, the exposed emulsion is first developed in a black and white developer to give a negative silver image but no dye, then the remaining grains are fogged and developed with a colour developer to give a positive dye image. The developed silver and any residual silver halide is removed by bleaching and fixing.

OTHER METHODS A monobath consists of a combined developer and fixer. A very active developer is needed so that development is complete before too much silver halide has dissolved. Since the process goes to completion in one step, the results are insensitive to the time of treatment. An adaptation of a monobath is used in image-transfer processing for black and white polaroid prints, the preparation of some printing plates, and other processes such as some photocopiers. The exposed film is developed in a monobath while being sandwiched against a receiver layer containing suitable nuclei on which the dissolved silver halide can be developed by the developer in solution. A normal negative image is obtained on the exposed photographic film, but on the receiver sheet, where the developed silver is derived from the (unexposed) silver halide grains that dissolve in the fix, a positive image is obtained. Again the process goes to completion and is insensitive to processing time.

One type of processing which is widely used, yet which many photographers rarely come across, is lith development. This is used in the graphic arts industry, particularly for reproduction in printing. Printing processes can only handle plain black or white tones; except for the gravure process, any half tones, or greys, have to be broken up into dots which look like a grey tone when seen from a distance. To make a negative of some subject for printing, therefore, a process is needed which can transform greys into either black or nothing at all: that is, with very high contrast.

The developer used for this process has very few free sulphite ions and is usually supplied as two solutions which are mixed just before use, to prevent oxidation. The lack of sulphite results in production of quinone by the first grains to be developed, which stimulates nearby grains to develop faster. In this way, development takes place rapidly in areas which have had a heavy exposure, and very slowly in the areas with less exposure. The development appears to spread out from the highlights of the picture, and is known as infectious development. The films used in graphic arts are designed with this in mind. A number of interesting special effects can be obtained in this way.

PROCESSING METHODS The mechanical procedures used to carry out the required processing steps depend on the scale of the operation. On a small scale, films and papers can be processed in dishes or held on frames and dipped into a succession of tanks for the required time. Roll films are usually wound on spirals and placed in a small light-tight box into which the processing solutions are poured.

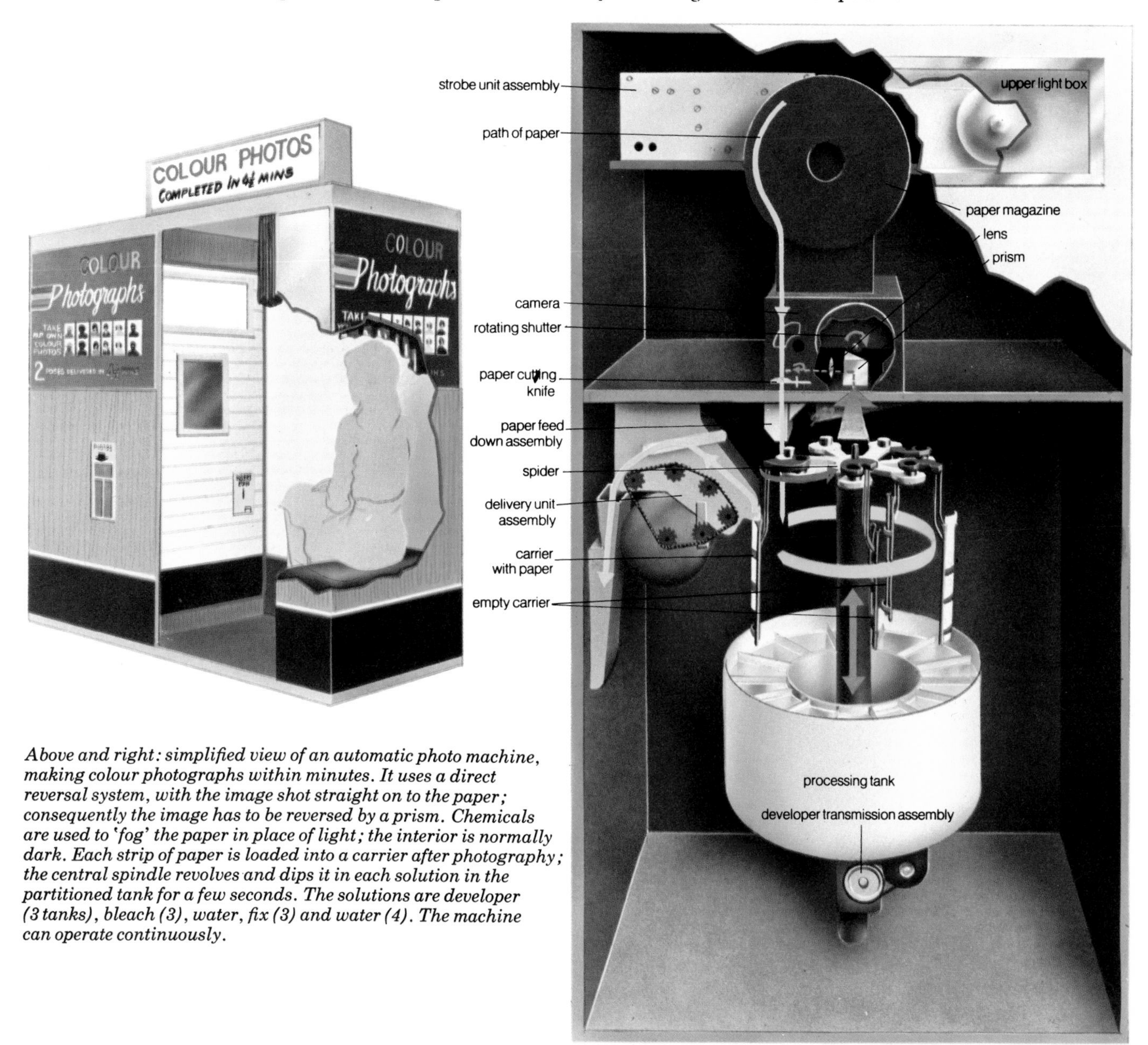

Above and right: simplified view of an automatic photo machine, making colour photographs within minutes. It uses a direct reversal system, with the image shot straight on to the paper; consequently the image has to be reversed by a prism. Chemicals are used to 'fog' the paper in place of light; the interior is normally dark. Each strip of paper is loaded into a carrier after photography; the central spindle revolves and dips it in each solution in the partitioned tank for a few seconds. The solutions are developer (3 tanks), bleach (3), water, fix (3) and water (4). The machine can operate continuously.

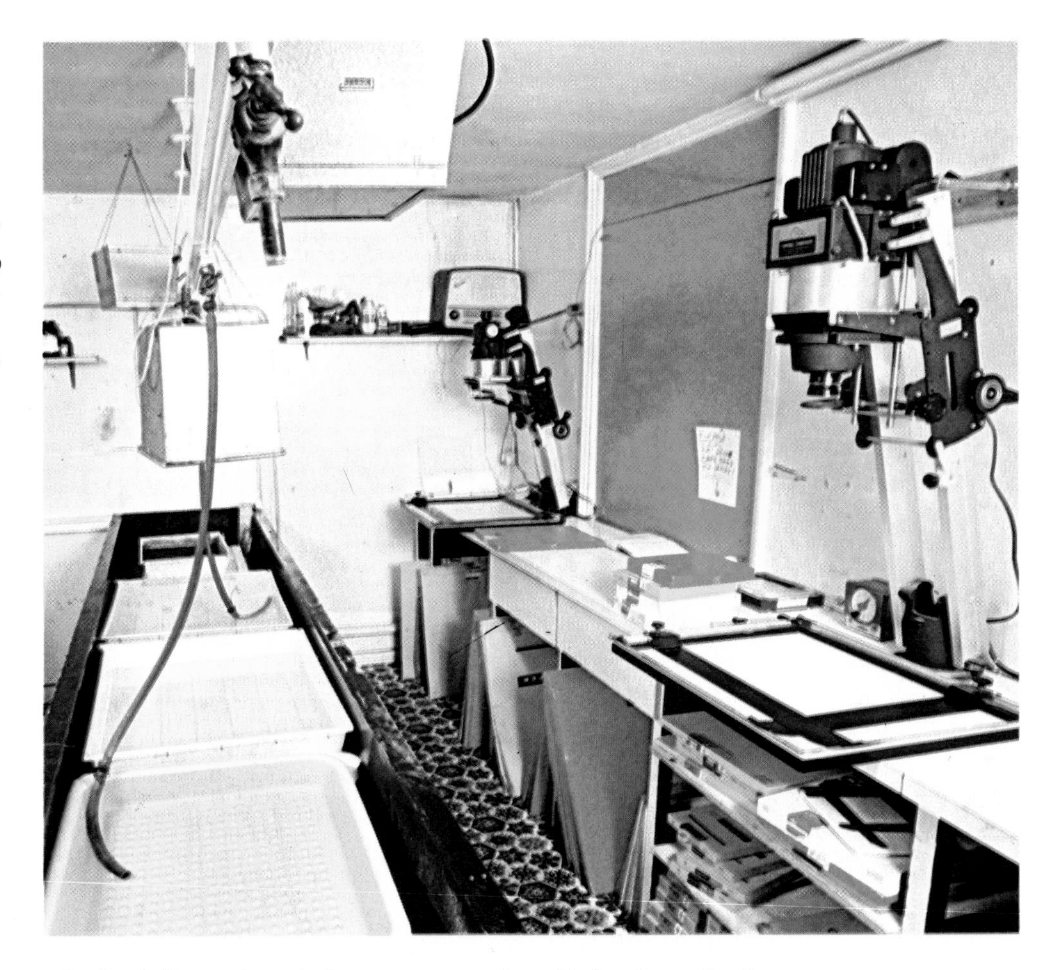

Right: the interior of a professional darkroom which produces individual prints of high quality. On the bench at right are two enlargers. When a print has been exposed (10 to 15 seconds) it is developed in the far dish (about 22 minutes), then transferred to running water in the next dish (about 30 seconds). The centre large dish contains fixer, the nearest one being water again. Fixing takes about 10 minutes, and the print is then washed for about 15 minutes. From the ceiling hang safelights which can be fitted with dark red or yellow filters which do not affect the paper. Small prints for family snapshots are not made in this way, but are produced in an automated plant on rolls of paper.

Processing on a large scale is fully automated. Photographic paper is processed in continuous rolls several hundred feet long and only cut up into separate prints when processing is complete; roll films are spliced together into long lengths. These long lengths of material are fed continuously on a system of pulleys in and out of a succession of deep tanks containing the processing solutions. The time spent on any processing step is controlled by the rate at which the strip moves and the number and depth of the tanks devoted to that step.

PLASTICS manufacture

The word 'plastic' comes from the Greek word *plastikos* meaning 'capable of being moulded', and it is the chief property of plastics that they are deformable and therefore easily made into almost any shape by processes such as moulding or extrusion. Plastics are composed of high molecular weight organic molecules, or polymers, made up of repeating units chemically linked together in the form of a chain or network. Plastics can conveniently be divided into two categories: semi-synthetic, in which the basic chain structure is derived from a natural product such as cellulose; and fully synthetic, in which the chain is built up chemically from small units, or monomers. The process of forming a polymer from its constituent monomers is called polymerization.

The first plastics to be manufactured commercially were semi-synthetic, and they were derived from the carbohydrate cellulose which was usually obtained from cotton waste. In 1862 the British chemist Alexander Parkes prepared a plastic material called 'Parkesine' which could be readily moulded and shaped. This was made by reacting cotton waste with a mixture of nitric and sulphuric acids to give a nitrocellulose compound which was then mixed with castor oil, a little camphor and a colouring material. Although Parkesine was easy to prepare on a small scale, it proved difficult to make in large quantities, and its industrial manufacture was not a success. In 1870 the American chemist John Wesley Hyatt prepared the first commercially successful plastic, Celluloid, which was similar to Parkesine but used camphor in place of the castor oil. The new material was used to make a wide variety of products including spectacle frames, combs, billiard balls, knife handles and photographic film. The first synthetic fibre, introduced in 1889, was an artificial silk made of nitrocellulose.

The chief drawback of these early plastics was that they were extremely flammable, which is not surprising when one considers that the main ingredient, nitrocellulose, is very closely related to the explosive guncotton; both are forms of nitrated cellulose. For this reason another ester of cellulose, cellulose acetate, is normally used nowadays in preference to cellulose

Above: the excellent electrical insulation properties of plastic are put to use in this elevating plastic bucket used by linesmen for repairs.

Right: washbasins heat-formed from a Perspex sheet.

nitrate for preparing cellulose-based plastics. These plastics are mainly used for making textiles. Regenerated cellulose, or Rayon, is a fibrous material composed of cellulose whose molecules have been shortened by dissolving and reprecipication.

SYNTHETIC PLASTICS The first fully synthetic plastics were made many years before they were manufactured on an industrial scale. In 1838 the Frenchman Regnault observed 'resinification' of vinyl chloride under the influence of sunlight but he did not appreciate the possibilities of the new material he had produced, now called polyvinyl chloride (PVC). (PVC is the plastic of which gramophone records are made.) The following year, the German chemist Simon reported the polymerization of the unsaturated hydrocarbon styrene to give polystyrene, but it was not for another nine decades that this material was made industrially. In 1909 Leo H Baekeland patented a resinous product he had obtained by reacting phenol with formaldehyde although the first phenol-formaldehyde resin had been discovered more than 30 years before by Adolf von Bayer

Above left: a field of potatoes is covered with polyethylene to retain heat and moisture.

Above: making rope from polypropylene, which is both strong and easily extruded.

in 1872. Baekeland's product, called Bakelite, was one of the first commercially successful synthetic plastics. Methyl methacrylate was first polymerized in 1877 by the German chemists Fittig and Paul, and the product they obtained, polymethylmethacrylate, was to form the basis of the plastic 'Perspex' ['Plexiglas'] which was introduced half a century later. The first experiments in polymerizing ethylene were conducted in 1879, but a lubricating oil and not a plastic was produced. Nowadays ethylene is polymerized industrially on an enormous scale to give the well known plastic polyethylene, or polythene.

Although celluloid was being manufactured commercially at the end of the 19th century, it was not until early this century that the first really significant advances were made in manufacturing techniques. Baekeland succeeded in developing a moulding technique for his phenol-formaldehyde resin Bakelite, and in 1919 Eichengrün was granted a patent for the injection moulding of cellulose-acetate. Nowadays injection moulding is an extremely important technique for shaping plastics, and without it many utensils and implements could not be made.

POLYSTYRENE In the 1920s much research work was done into the chemical structure of plastics, particularly by the German chemist Hermann Staudinger who was awarded the Nobel Prize for chemistry in 1953 in recognition of his work. Staudinger's research led to the development of new plastics on a more rational basis, and in 1930 commercial production of polystyrene, one of the most important of today's plastics, was launched by the German company IG Farbenindustrie. The raw materials for polystyrene production are ethylene and benzene, which are first reacted together to give styrene. The styrene is then polymerized to give polystyrene, whose molecules consist of chains of styrene units.

There are various other important plastics which are related to polystyrene, for example, impact resistant polystyrene (a mixture of polystyrene and rubber), ABS (a polymerized mixture of styrene, acrylonitrile and nitrile rubber), SAN (a polymerized mixture of styrene and acrylonitrile) and ASA (a polymere of styrene with a methacrylate ester). Polymers composed of more than one type of monomer, for example SAN, are usually called copolymers. Polystyrene plastics are used to make mouldings and in various electrical components. Expanded polystyrene foams, such as 'Styropor', are used as heat and sound insulation materials and in packaging to protect against mechanical shock.

PVC Another important modern plastic is polyvinylchloride (PVC) which was first produced commercially in Germany in 1931. The raw materials for PVC production are ethylene and chlorine which are reacted to give vinyl chloride. This is polymerized to give PVC.

PVC is used to make a wide variety of moulded products, and it is often mixed with a plasticizer which softens it and makes it more flexible. Plastic pipes and gutters, and flexible plastic sheets are frequently made of PVC. Copolymers of vinyl chloride with vinyl acetate or vinylidene chloride are also important commercial plastics.

POLYOLEFINS In 1936 the British company ICI succeeded in making the first aliphatic polyolefin plastic, polyethylene. Olefins are hydrocarbons which have one or more double bonds, and ethylene is the simplest member of the group. It is polymerized at about 200°C (392°F) and a pressure of more than 1000 times atmospheric pressure to give molecules of polyethylene. The polyethylene produced in this way is called low density polyethylene which is a tough, flexible material. High density polyethylene is more rigid than the low density variety and is made by polymerizing ethylene at much lower pressure and in the presence of a catalyst. The difference between the two types is that the molecules of high density polyethylene are straight whereas the molecules of low density polyethylene are branched. Another important polyolefin plastic is polypropylene, made by polymerizing propylene.

Polyethylene and polypropylene are used to make plastic bottles and other containers, films for packaging, pipes for plumbing and many other applications.

Another plastic which is related to the polyolefin plastics is polytetrafluoroethylene (PTFE) made by polymerizing tetrafluoroethylene. It is used where heat resistance and low surface friction are important. Non-stick coatings on domestic cooking utensils are generally made of PTFE.

POLYAMIDES The first polyamide plastic was prepared in 1934 by the American chemist W H Carothers, although it was not until 1937 that production was started on a commercial scale. The new plastic was called nylon. The chief raw materials used in the production of nylon are benzene and butadiene. Various intermediates, particularly caprolactam, are produced before the final product is obtained. Polyamides are chiefly used for making moulded articles and textiles.

POLYESTERS These plastics are made by reacting an organic acid having at least two acid groups with an alcohol having at least two alcohol groups. Polyesters are used to make high quality plastic films as well as synthetic fibres, such as Terylene [Dacron].

POLYURETHANES Polyurethane plastics are widely used today, particularly in the form of flexible or rigid foams. Flexible polyurethane foam is used as an upholstery material, and the rigid foam is commonly used as a heat insulating material. Polyurethanes are also used in some paint compositions to impart surface hardness to the paint coating. They can be made by reacting an isocyanate having at least two isocyanate groups with an alcohol having at least two hydroxy groups.

Most of the plastics discussed above are thermoplastics, which means that they will soften when heated and harden again when cooled. Some plastics, however, such as Bakelite, are thermosetting, in other words they cannot be softened by heating without destroying the chemical structure. Another example of a thermosetting resin is the melamine formaldehyde type such as Formica. Some adhesive compositions, particularly epoxy adhesives such as 'Araldite', are also based on thermosetting plastics. Another common group of thermosetting plastics consists of the alkyd resins which are used in some paint compositions.

Above left: strands of nylon-6 being drawn from the cooling trough at the base of a polymerization reactor (the picture exaggerates their thickness).

Left: a bed of urea formaldehyde foam was used to bring this aircraft to rest from 90km/h (60 mph) without damage. Such beds are now used at airports.

PRINTING

Apart from the obvious books, magazines and newspapers, the products of the printing industry are many and diverse. They include posters, banknotes, telephone directories, postage stamps, record sleeves, wallpapers, cartons, plastic containers and many other forms of packaging. Today, even electronic circuits and working surfaces for kitchens are printed.

As the main vehicle for the conveyance of ideas during the last five hundred years the printing press has had its influence on politics and government, on literature and education, on business and economic affairs and on the development of society as a whole. In turn the demand for printed matter in new forms and in greatly increasing quantities has led to revolutionary changes in the technology of printing.

Although the modern printing industry is generally considered to have begun with the invention of movable type in the mid-fifteenth century, the Chinese had been printing on paper many hundreds of years earlier. Paper originated in China in about 105 AD, when papyrus and parchment were being used in Mediterranean countries. Printing from wood blocks was done in China and Japan from the sixth century onwards, and in 767 AD the Empress Shiyautoku is said to have ordered a million copies of a Sanskrit charm in Chinese characters for distribution to her people. The invention of movable type has also been traced back to China, where Pi Sheng in 1042 AD used clay for moulding type. Nearly 300 years later, Wang Chen introduced wooden types which were less easily broken. These types did not lead to printing in the Western sense because typesetting would have been difficult, since there are about 40,000 characters in the Chinese language.

The art of papermaking did not spread to Europe until about 1157 when Jean Montgolfier escaped from the Saracens and returned to France to set up a paper mill. It is said that he learned the art while working as a slave in a paper mill during the second Crusade. For three centuries after paper became known in Europe books were still laboriously produced in very small numbers and their use was confined to the churches and monasteries. It was only in the fifteenth century that the invention of methods of printing books allowed the spread of knowledge to pave the way for learning and culture.

In 1440 Johann Gutenberg, a Mainz goldsmith, began experimenting with printing when he was a political refugee in Strasbourg. He later returned to Mainz and by 1450 he had developed his invention to the point where it could be exploited commercially. In partnership with a lawyer called Johannes Fust, who advanced the money, he commenced casting metal type. Gutenberg soon ran into serious financial

Left: a Swiss woodcut of 1586 showing a printing press. The paper is placed in the tympan while the type is inked by means of pads. The impression will be made by a single horizontal pull of the lever, which is attached to a coarse screw thread. In the background, compositors set the type. For a large book, so much type would be used that the same type would be used several times over, each page of type being broken up after printing.

Above right: an 1832 engraving showing typefounders.

Right: schematic layout of the typefounding process. The artist's drawing is projected to make a large original, which is then reduced on to wax by a pantograph. A copper shell is made from the wax blank, and a steel master punch cut from this with a vertical pantograph. This makes the matrices or moulds.

trouble and in 1455 Fust terminated the agreement, the bulk of Gutenberg's types and presses going to Peter Schöffer, who was in Fust's service and later married his daughter. There are rival claims for the invention of printing on behalf of a Dutchman, Laurens Coster, but information about him is very scarce and none of his work remains.

The only major work that can confidently be called a product of Gutenberg's own workshop was the 42-line Bible published in 1456. This was printed a page at a time in an edition of two hundred copies of which about thirty were printed on vellum. Its quality was such as to rival the best manuscripts. Gutenberg's achievement lies not merely in his invention of a practical method of printing books, but in the standards of technical perfection established in his workshop, which were not to be surpassed until the nineteenth century. In today's world of specialization it is also worth noting that these early printers combined the arts of punch cutting (making the 'patterns'), matrix fitting (making moulds from the punches), type founding, ink making, printing and publishing.

Within fifteen years of Gutenberg's death in 1468, printing presses had been set up in every country in western Europe, most of the printers being German nationals. William Caxton, the first English printer, learned the trade in Cologne, and after a period of

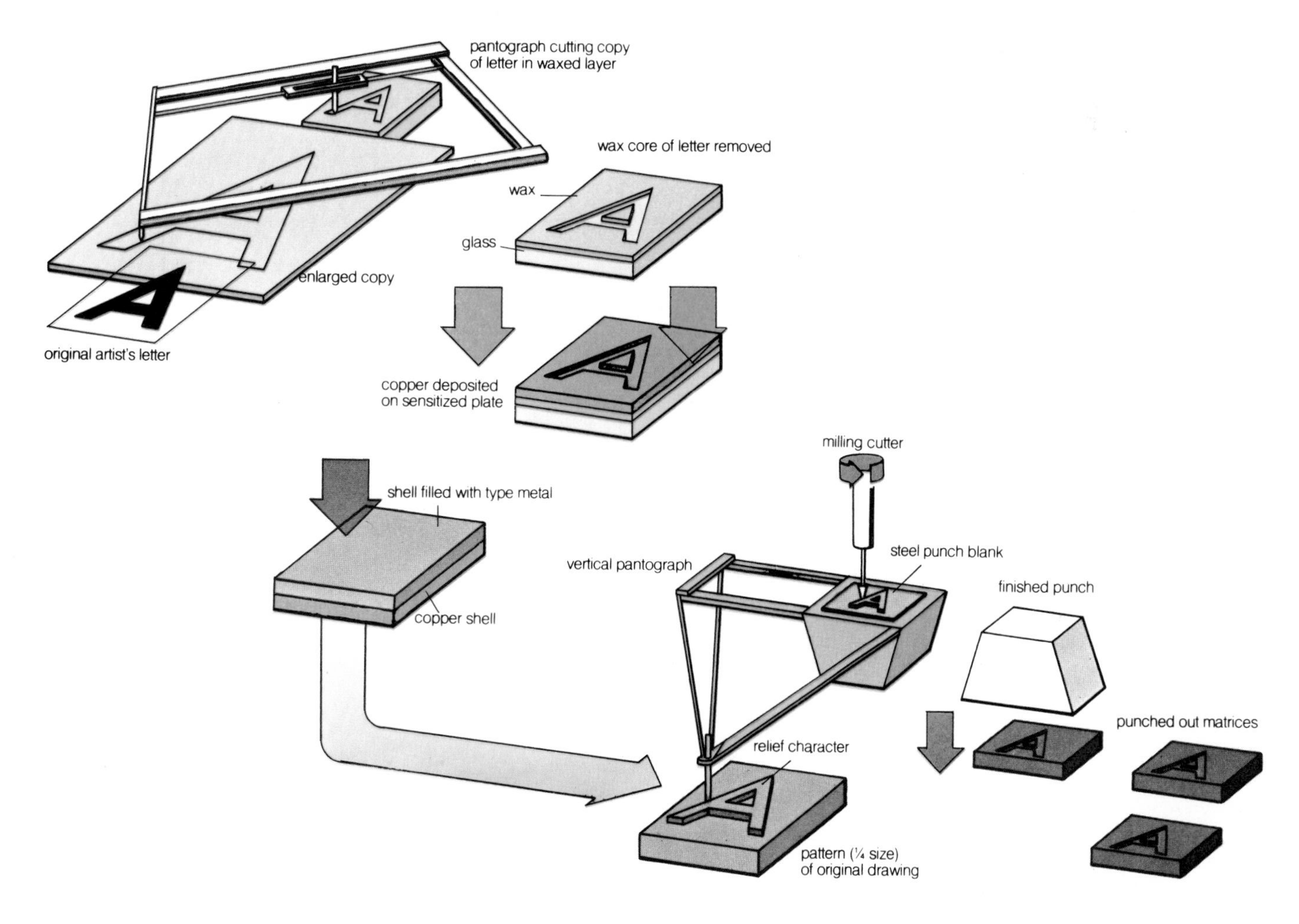

printing in Bruges, he returned to England in 1476 to set up his own press in Westminster.

The early printing presses of the fifteenth century were of the wooden screw type and were little more than adaptations of the wine presses and bookbinding presses of that time, printing one sheet at a time. They were levered by hand and so demanded a great deal of muscle power. The area of the *forme*, in which the type is held in the press, was very small because of the limited force applied and repeated readjustments of the sheets were necessary. In the early sixteenth century a metal screw was introduced and the press improved by the use of a sliding bed (on which the forme lay, face up, to be inked and then slid under the platen, the plate which applied the pressure), but no basic changes in the technique of printing were to take place for three hundred years. The beginning of the nineteenth century saw a period of technical advance which has continued to the present day.

In 1800 Earl Stanhope invented an all-iron press with a screw and a system of levers which produced improved impression on a larger platen area. This made it possible to print a large forme in one pull, whereas the wooden press required two pulls, each one covering only half the area. Hand presses showing further improvements over the Stanhope press were later introduced, including George Clymer's Columbian Press (Philadelphia, 1816), which was the first

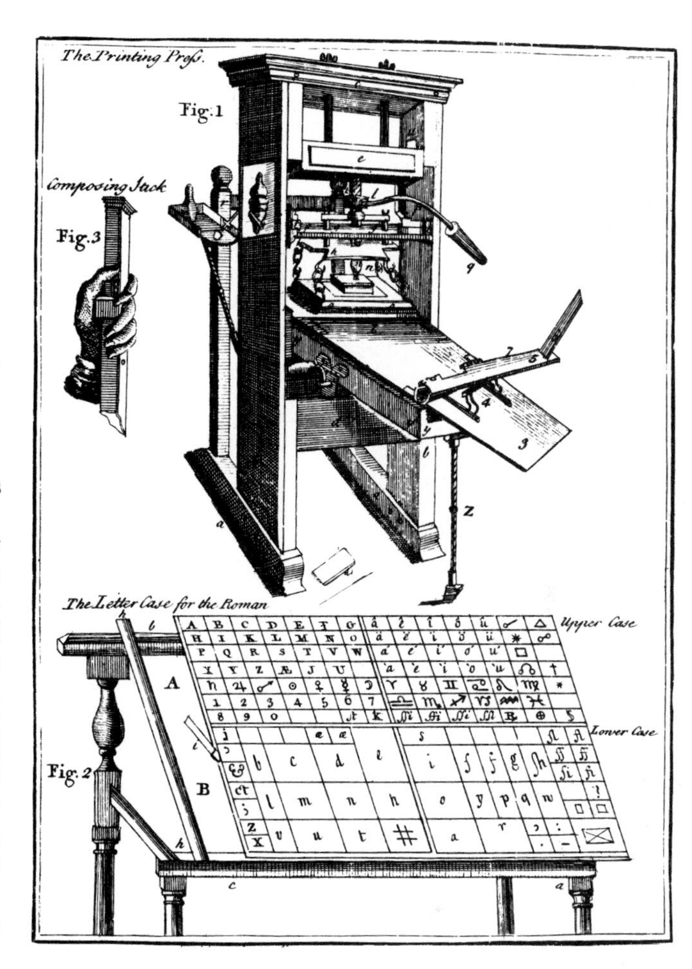

Left: in this late 18th century view of a printing press, not much has changed since the earlier picture on p. 130. Typefounders are at work at the rear; more type was always needed through breakages.

Right: a printing press of 1747, with a type case including many symbols rarely found today. It does, however, show why small letters are known as 'lower case'. By contrast, the modern job case (below left) has the lower case on the left, upper (capitals) on the right and figures, spaces and punctuation in the middle. Different typefaces are stacked below.

Below right: an early hand operated cylinder press for newspapers, made by Hoe of New York. The paper was fed round the large drum, which rotated as the inked type slid beneath it.

iron press without a screw, and Cope's Albion Press (1830).

A German named Freidrich König, intrigued by Watt's invention of the steam engine, set out to design a printing press powered by steam. Unable to find support for the idea on the continent he came to England in 1806 and developed the first mechanized platen press, capable of 400 impressions an hour. In 1811 König patented a steam-powered cylinder press, in which a horizontal printing forme moved back and forward under a revolving cylinder, printing over 800 sheets an hour. Two of these presses were ordered by *The Times* newspaper, and the first paper produced on the new press was that of 29 November 1814. Two years later König produced the first 'perfecting' press, which allowed both sides of a sheet of paper to be printed in one pass of the machine.

Above: screen printing is widely used for posters and other jobs where a short run is required. This is nothing to do with the screen of halftone dots used for printing illustrations; the image is transferred to fabric in such a way that it is partly impervious and partly porous. Ink is forced through the porous parts on to the paper.

Right: letterpress machines.

Below right: a one-piece plastic letterpress plate.

Letterpress remains the only process capable of printing from individual type characters. It is still widely used for work in which type matter predominates and where the ability to make late corrections, if necessary on the press, is of importance. Although offset lithography now accounts for a larger volume of printed work, letterpress is still extensively used for newspapers, books and general printing. A letterpress printing surface may consist of type alone, or the type may be combined with photo-engraved plates which print illustrations. There are several thousand different typefaces, although only a fraction of these are in general use. Each family of type follows one basic design but generally provides characters which are upright (roman) or slanted (italic), variations in weight between light and bold faces, and may include expanded and contracted versions. Most typefaces are produced in a range of different sizes, the unit of measurement being the 'point' which is approximately equal to 1/72 of an inch. A set of type of the same style and size is called a fount. Types are cast from alloys containing lead, tin and antimony, the proportion of each metal varying slightly depending on the casting method.

TYPESETTING Type is still set by hand for a wide variety of letterpress work. The compositor selects the

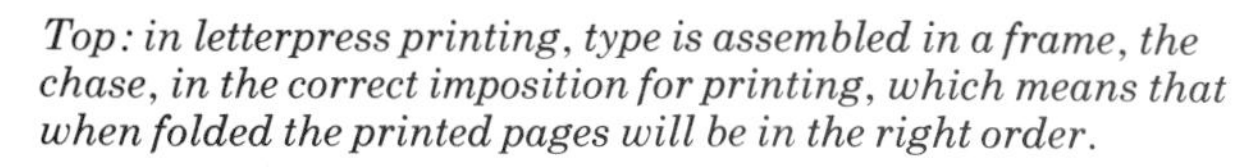

Top: in letterpress printing, type is assembled in a frame, the chase, in the correct imposition for printing, which means that when folded the printed pages will be in the right order.

Above: a flat metal stereotype with the intermediate or mat made of flong, a paper pulp board with a coated surface.

Right: various methods of obtaining an impression in letterpress printing. In the platen press (top) the forme is held vertically and the platen closes, pressing the paper against the image. In the flatbed press (centre) the forme is inked and moves under the impression cylinder while a sheet of paper is fed between them. The fastest press (bottom) uses the rotary principle and the printing surface is curved around a cylinder.

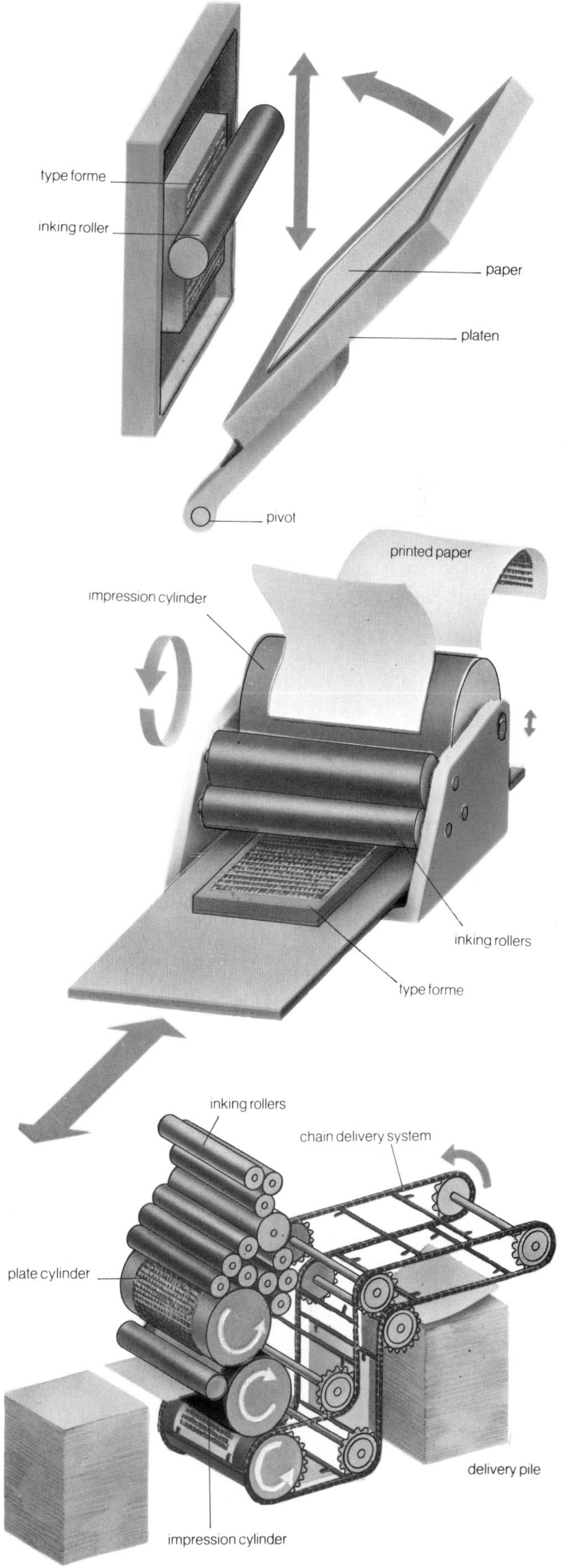

Above: a small proofing press.

Below: in Linotype setting (left) the operator depresses the keys, releasing the matrices on to a belt which takes them to the assembler and caster. In Monotype (right) compressed air controls draw rods and hence the position of the matrix case over the mould.

required letters from the type case, together with necessary spaces, and assembles them in a holder called the stick. When the stick is full, the lines of type are transferred to a flat metal tray, the galley, where they are separated into page lengths. After the page has been made up, it is tied with cord so that it can be safely moved to the proofing press for a galley proof to be produced.

After galley proofs have been taken and any corrections made, the pages of type are assembled inside a metal frame called the chase. The pages must be arranged in such a way that when they are printed on to a sheet of paper, and the paper folded, they will be in the right order in a book or magazine. This is called imposition. Spacing material, called furniture, is then added to hold the pages in the correct position and the whole assembly is locked up with the aid of quoins. These are devices which, when turned with a key, exert lateral pressure on the furniture on either side of them. The locked-up assembly, known as the forme, is now ready for further proofing and finally printing.

LINOTYPE AND MONOTYPE Typesetting by hand is obviously too slow and expensive when large amounts of text are to be printed. In order to meet the demand for the modern mass production of newspapers, magazines, and books, methods of mechanized typesetting have been explored over a long period of time and two distinct systems have developed. In the first of these, line composition, for example Linotype and Intertype, characters are cast in slugs. In the second, single letter composition, for example Monotype, type

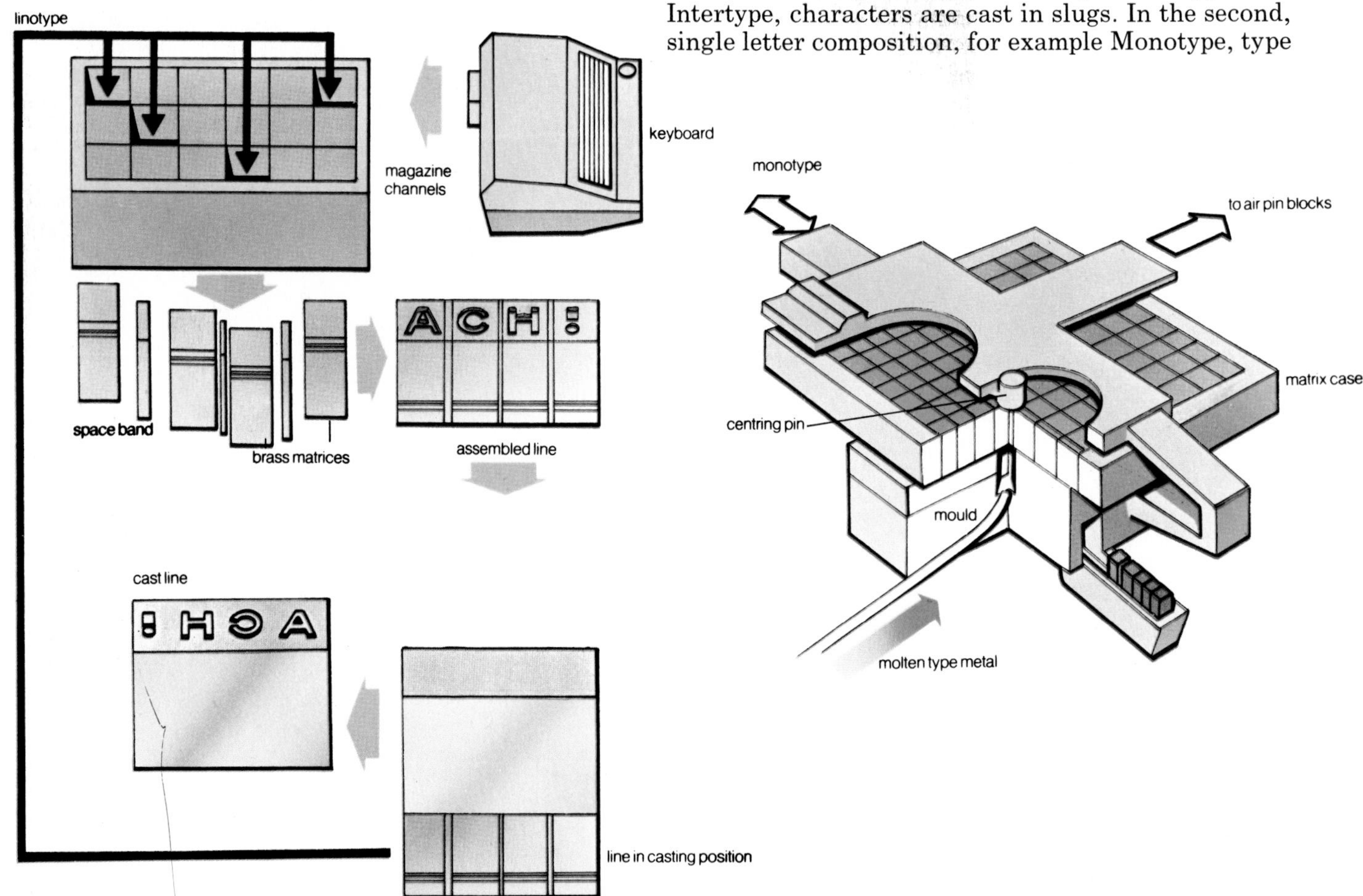

Left: with the copy pinned in front of him, the operator keys the words on a line composing machine to release the matrices.

Below: a Monotype keyboard. The information keyed in is recorded on a paper spool.

Right: one of the earliest line composing machines; they were introduced in the 1880s.

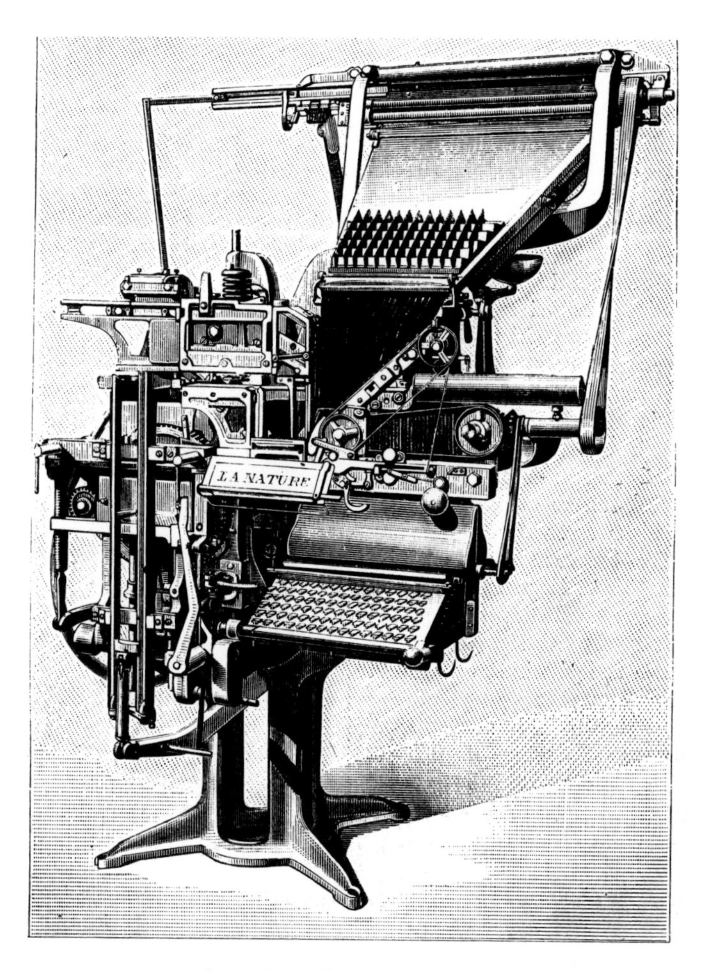

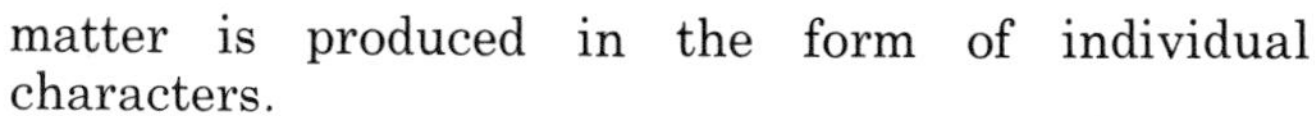

matter is produced in the form of individual characters.

Line composing machines are single units which combine the functions of keyboarding and casting. The keyboard consists of ninety keys arranged in six rows of fifteen. The operator reads the copy (text) and depresses the keys accordingly. Each keystroke releases a matrix from a magazine. The matrices are thin rectangles of brass, with a character engraved in both a roman and an italic form on one of the longest sides, and a series of teeth on their upper edges used to correctly distribute them in the magazine. The matrices are held in magazines at the top of the machine, each character being held in a separate channel. On being released by the keystroke, the matrix falls on to a moving belt which carries it into an assembly box. A wedge-shaped spaceband is inserted between words by depressing the spaceband key on the keyboard.

When the end of a word is reached, completing a line which fills or almost fills the measure being set, the operator presses a lever which raises the assembled matrices and spacebands to the first elevator. From this point on, the casting operation is carried out automatically, and the operator is free to start tapping out the next line. After the line is in position in front of the mould, the spacebands are elevated and their wedge shape enables them to take up any slack in the measure, increasing each word space by an equal amount to produce justified lines (lines all the same length).

A crucible containing hot molten alloy is now closed against the back of the mould wheel and a piston-shaped plunger forces a jet of molten alloy into the slot in the mould and against the matrices to produce a line of type or slug. The mould wheel, with the slug in position, now turns to allow the back knife to trim the bottom of the slug to type height (0.918 inch, 23.317 mm), an ejector blade pushes the slug between parallel knives which trim the back and front to the correct point size, and finally the slug drops into the delivery gallery. As the slug is being trimmed and ejected, the matrices are carried up to the top of the machine where they travel along a vee belt and drop into the appropriate channel of the magazine, ready for use in casting subsequent lines.

Linotype and Intertype machines are the most widely used line composing machines and although they differ in their detailed construction, their working principles are basically similar. A range of models is available designed for different kinds of work. Although each face (design) and size of type requires a separate magazine of matrices, machines may hold several magazines and 'mixer' machines are available which make it possible to use different faces within the same lines.

In recent years slug casting machines have been adapted for automatic operation from paper tape. The tape may be punched locally or it could be transmitted by land line from a central point to several associated printing plants, for example producing local editions of a national newspaper.

Line composing machines are ideally suited for newspaper and magazine work, which demand the rapid production of a large amount of text of a uniform size. They also offer a fair amount of versatility and they are used by many jobbing printers for a wide variety of work.

Single letter composition involves the use of two separate units, a keyboard and a caster. The product of the keyboard is a punched paper spool, which is subsequently used to control the operation of a casting machine producing composed matter, or alternatively type for the case.

A Monotype keyboard has two sets of keybanks mounted side by side, each following the standard typewriter layout. When the operator taps a key, compressed air causes punches to perforate holes in certain positions across the width of the paper spool. The machine has thirty-one punches, and each character is represented by a particular arrangement of between one and four holes. When a key is struck, a pointer moves along an *em* scale (an em is $\frac{1}{6}$ inch, 0.4233 cm) above the keyboard, and the operator is able to see the amount of space that has been used up. On releasing a key the paper ribbon advances $\frac{1}{8}$ inch (0.3175 cm) ready for the next set of perforations to be made. When the setting of a line is near to the limit of the measure, a warning bell rings, and the operator completes or breaks a word according to the amount of space left. He then depresses two justification keys, the numbers of which are indicated to him by referring to a justification drum positioned above the em scale. A key button returns the scale indicator and the operator continues setting until the line counter shows that there will be sufficient type to fill a galley.

The paper spool is now transferred to the casting machine, being fed over a line of thirty-one air pipes corresponding to the thirty-one punches on the keyboard. Compressed air passing through the perforations for a particular character raises pin blocks which control the position of the matrix case over the orifice of the mould. When the matrix case is in position, a tapered centring pin moves down into a cone-shaped hole at the back of the matrix holding it firmly against the mould. Molten type alloy is now forced into the mould, which is capped by the selected matrix to form the type character. The letters are cast one by one, held in a slide until the line is completed and then the whole line is pushed out onto a galley.

The paper spool is fed into the caster in the reverse direction from that in which it was keyed, so that the last line to be tapped is the first to be cast. For any given line, this means that the caster first 'reads' the justification perforations, which position the wedges so that the correct amount of space between each word is determined, before the casting of the line is commenced.

The matrices are made of phosphor bronze and each one measures 0.2 inch (0.5 cm) square. They are arranged in rows in the case, the number of rows depending on the model of machine. Early models had fifteen rows of fifteen matrices (225 positions); later machines had seventeen rows of fifteen (255) and unit shift machines have seventeen rows of sixteen matrices (272).

The Monotype Super-caster is of a similar size to an ordinary composition caster but is specially designed for casting type for the case from $4\frac{1}{4}$ to 72 point (72pt=one inch) and also for casting leads (spaces between lines), rules, borders, and spacing material

Top: a Monophoto operator inserts a matrix frame in the 'caster'. This small frame holds the master negative of all the characters in the typeface specified. The machine itself can be set for various magnifications. Here the end product is not the conventional three-dimensional piece of type but a two-dimensional photo image.

Above: an Intertype machine consists of two separate units, a keyboard and a caster. The keyboard produces a perforated paper spool which controls the caster. Here metal is loaded into a caster.

The Monotype system of composition has proved its value for a wide range of work but is particularly suitable for book production involving symbols and numerals. Compared to line composition, it has the advantage that corrections can be made without going back to the typesetting.

In recent years a firm trend has been established towards letter assembly on photographic film rather than in hot metal type.

PHOTOTYPESETTING Phototypesetting (also called photocomposition or filmsetting) is a process for the automatic selection and projection on to photographic film of text ready for the various printing processes. The characters are most commonly stored in the machine in a manner similar to slides in a projector and the machine receives instructions such as particulars of the size and style of the characters, the order in which each character and space must be exposed, the length of the line of type to be set, the vertical space between each line, and many other factors, from a punched paper tape, 7 or 9 track magnetic tape, or magnetic disks.

The process of making each line the same length is known as *justification*. Another difference between print and typescript is that most typewriters allocate a uniform width to each character, irrespective of whether the letter is as wide as this capital W or as narrow as this l. In print, however, much careful attention is paid to character fit, which improves the appearance of the text matter but requires that each character is given only as much space as its width demands. Justification and character fit are, in fact, common to the hot metal typesetting method which uses cast metal type, but are mentioned here because in phototypesetting one of the most important items of data to be fed into the machine is that which specifies the 'set feed' for each character.

There are only two methods of arranging type matter across the page: either the source remains stationary and the page is moved (as with the carriage of most typewriters) or a system must be devised to project each successive image in its correct position along the line. Various manufacturers of filmsetting machines have their own methods of doing this, but most use one of two systems of mirrors.

Both systems are substantially similar, the main differences being the light source and the method of storing, selecting and presenting the characters. In the first type, 400 film matrices (each carrying a character image) are arranged in a square grid of 20×20 with each character located by a simple two-part address which defines column and row as in a map reference. The matrix case is moved simultaneously in both directions to locate the required matrix beneath a single constant light source which is controlled by a mechanical shutter. The beam of light carrying the image passes through a condenser lens, is turned through 90° by an angled mirror, projected through a zoom lens (which controls character size and focus) and then to a system of moving set feed mirrors which are used to expose the images of the characters in their

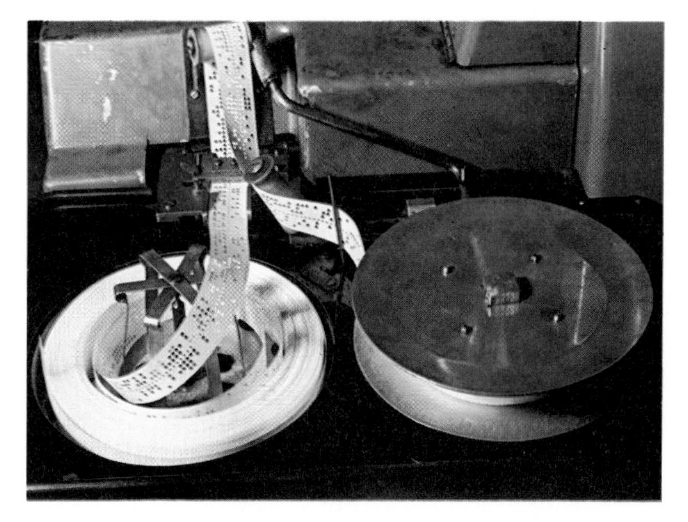

Above: a paper tape reader and the punched tape of instructions. The punch has a typewriter keyboard plus extra keys for instructions to the system.

Below: matrix discs; the matrices, similar to photographic slides, each of one character, are around the edge.

Bottom: An alternative method of storing the matrices is to arrange them on a grid. This one holds 400.

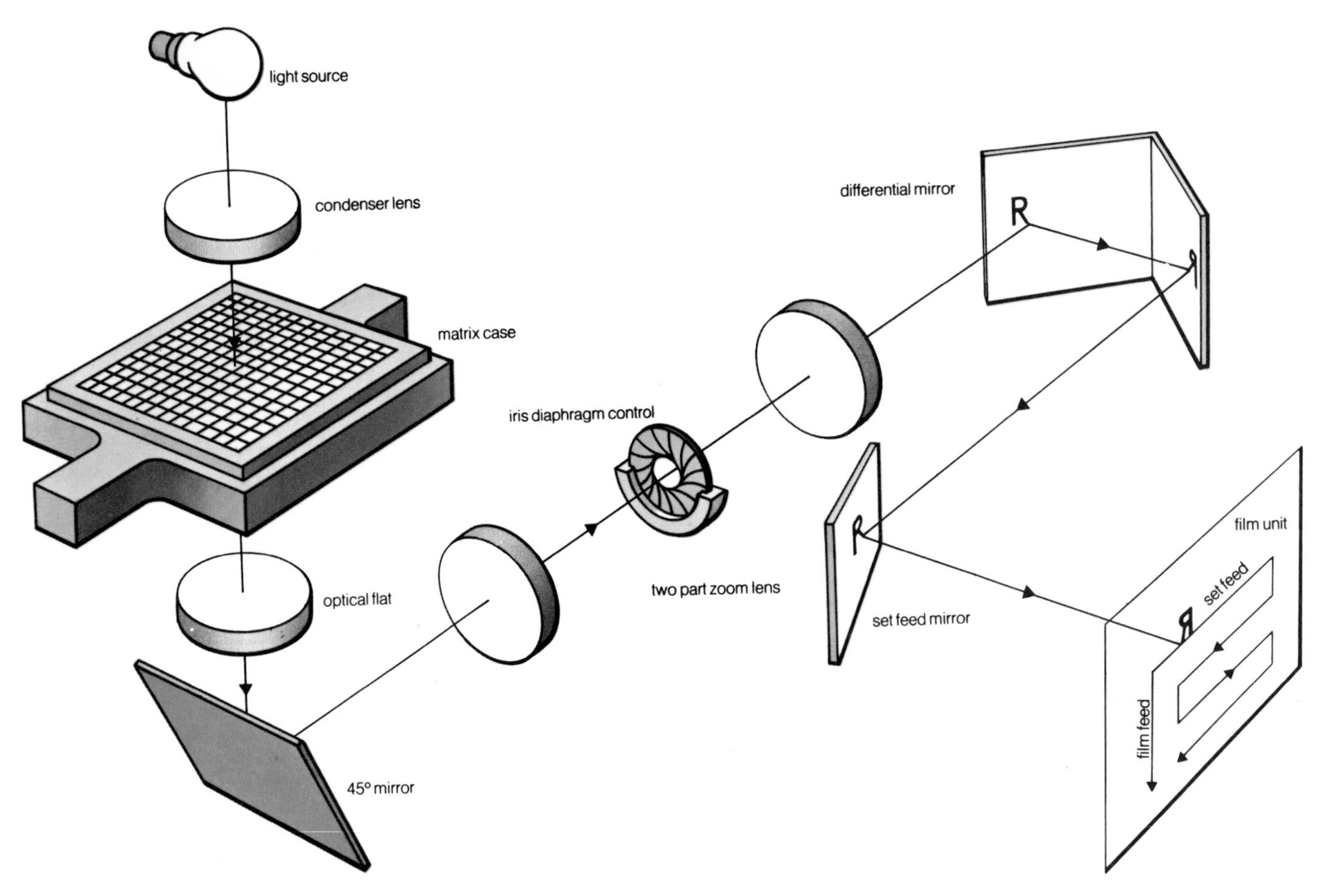

correct sequence and spacing across the page.

In the second system, the film matrices are disposed around the periphery of four discs. Characters are selected by rotating the discs in either direction to align them accurately in the optical path of the light source. Each disc has its own xenon flash tube and the four projection paths are merged by means of a multi-mirror block of semi-reflecting surfaces so that the character images on any disc can be centralized on a common optical target. The indexing and arrest of the discs, to locate the selected matrix prior to exposure, are carried out at high speed, which is further improved by the way in which the characters are grouped around the four discs. Characters are assessed for their frequency of occurrence (e being the most common character, a the second, t the third and so on) in conformity with a system which makes it unlikely that two consecutive characters will be exposed from the same disc. By such means matrices can be selected and presented for exposure in advance, and the matrix discs operated in sequence. The image then passes through sizing lenses to set feed mirrors in a manner similar to that described for the grid system.

These two systems share a common characteristic, namely, that the character is exposed on the film from a stationary matrix. This system claims the advantage of offering very good typographical quality, though the speed of setting is not so high as that for machines which do not hold the matrix still during exposure.

The second system, which uses the four discs, will expose some 35 characters per second, which represents the practical limit if the disc is to be arrested and the matrix kept still at the moment of exposure.

Where quality may be of less significance than the speed at which the setting must be performed, the matrices can be arranged on glass discs which spin at very high speeds and exposure depends on the very accurate aiming of a stroboscopic flash which fires and 'freezes' the selected character as it passes through the optical path. Machines using this system are capable of setting up to 75 characters per second.

Yet another system dispenses with a complete image of the characters but instead stores in a computer memory masses of individual terms of data which describe the configuration of parts of the letters. Thus, the image to be projected never, in fact, exists in complete physical form, but is generated from a great number of discrete parts. The data are used to control an electron beam, which, by its scanning action, 'paints' the character on the screen of a cathode ray tube (crt) in the same way as a picture is composed in a television receiver, and the character image is recorded on to film.

Other 'crt' machines use a combination of the systems previously described in that they employ an electron beam to scan film masters in a grid type matrix case and project the image through an optical system on to the film.

On the completion of a line the filmsetter receives a line feed signal, in response to which the film is 'wound-on' an accurately specified distance to produce the required inter-line spacing. In normal text matter this would be constant from line to line, but subject to variations to divide paragraphs, leave space for sub-headings, chapter titles, and footnotes, for example, which may appear in different type styles. When a complete section has been exposed on the photosensitive material which, despite it being referred to as film, can also be opaque paper, the cassette in which it is contained is removed from the machine and processed.

The final product is then arranged according to the design of the page with text matter, headlines, chapter titles, page numbers and illustrations in the positions allocated by the designer. The result is camera copy from which printing blocks or lithographic plates, for example, are produced.

GRAVURE Gravure is one of the major commercial printing processes and can be used successfully on papers of qualities ranging from newsprint to fine art.

Top left: film matrices in a grid on a matrix case are moved beneath the light source so that the light shines through one character at a time in the correct order. The images are projected through the lens and mirror system on to the film. The machine is programmed like a small computer by a 31-channel tape loaded into its memory unit. Another tape controls character selection. The matrix case is moved by two hydraulic rams.

Left: a machine for preparing punched tapes. The central typewriter keyboard is flanked by keyboards for instruction code punching.

Right: proofing a gravure print. A colour proof, or sample copy, is made and inspected and any defects rectified on the copper cylinder by hand.

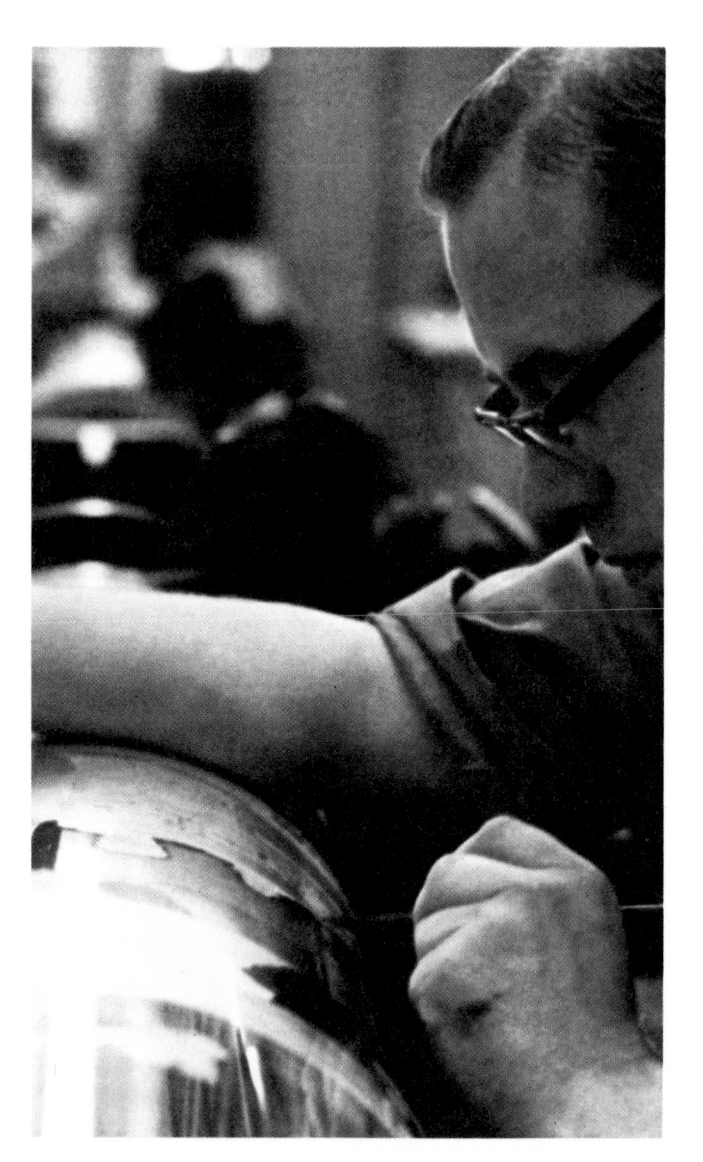

It is used mainly for printing magazines and packaging, but other applications include the production of decorative laminates such as wood grain effects, floor tiles, wallpapers, postage and trading stamps, and fine art reproductions. Gravure is an intaglio process; that is, the ink is transferred to the paper from very small cells which are recessed into the printing cylinder. Other engraving and etching processes are also called intaglio.

Because the cell depths (and in some processes, the cross-sectional area) can be varied, different amounts of ink may be transferred at different points, allowing subtle variations of tone such as are seen on a single colour postage stamp. The depth of a gravure cell may vary from about 0.001 mm in the highlight areas, to 0.4 mm in the shadows.

There are three main types of cell structure, depending on whether the surface is prepared by conventional gravure, invert halftone or electromechanical engraving methods. In conventional gravure the image is broken up into tiny square cells all of the same area but varying in depth. Invert halftone processes are characterized by varying cell cross-sectional areas; depths may either be kept constant or also varied. This is particularly useful for colour reproduction. In electromechanical engraving, signals from a scanning head, which is moved systematically across the photographic subject to be reproduced, are used to control the movement of a diamond stylus engraving head which cuts out cells in the shape of inverted pyramids with varying depths and areas.

Gravure printing surfaces are normally made of highly polished copper, which is often deposited as a thin skin by electroplating on to solid steel cylinders. For hard wear on especially long runs this may be chromium faced. Initially the material to be printed must be photographed to provide negatives of high contrast for line work such as text, but only medium contrast for continuous tones as encountered, for example, in a photograph of a scene. If necessary the negative may be retouched to improve the quality of the final job.

Positives and not negatives of both pictures and text are used for producing the image on the cylinders or plates. To do this carbon tissue is used, which is a

Below: a printing cylinder, chromium plated for long service, is furnished with ink on one side of a rotary gravure web-fed press. The surface is wiped clean by the doctor blade, leaving ink only in the cells. The web of paper, travelling at speed, comes into contact with the cylinder and the ink is transferred to the paper.

Near right: a photomicrograph shows the cells in detail. Deeper etching holds more ink and deepens the colour.

Far right: the copper-plated printing cylinder revolves slowly in the etching bath; conditions are controlled to give the correct depth of cells.

light-sensitized paper coated with a mixture of pigment and gelatin. The carbon tissue is first screened, to divide the printing area into small cells, by covering it with a gravure screen (composed of tiny opaque squares surrounded by clear lines) and by exposing it to light in a contact vacuum frame. The ruling of the screen is usually 60 lines per centimetre (giving 3600 cells per square centimetre), but may be up to 160 lines per cm. These rules provide the thin walls of the cells. Next the positives, mounted on a glass plate and in contact with the tissue, are exposed to diffused light. Where the tones are lighter the light passes through the positive freely and the gelatin on the carbon tissue becomes harder than where light reaches it weakly, from darker areas of the positive, while the screened lines remain the hardest.

Now the carbon tissue is ready to be mounted on to the cylinder. After mounting, the backing paper is removed and the cylinder developed in warm water to wash away any soluble (unhardened) gelatin, leaving the hardened gelatin, which is called the resist. This process is followed by drying and the painting out, with a bitumen base varnish, of any areas such as edges to be protected from etchant. Ferric chloride is used for etching and first penetrates the gelatin in the thinnest areas where the printed tones will be darkest, and lastly the highlights where the gelatin is thickest. Etching times vary from ten to twenty minutes and various factors must be controlled, such as the strength of the ferric chloride solution and the temperature of the bath. After etching is completed, the bitumen varnish and any residual gelatin are removed and the cylinder or plate washed and dried. It is now ready for proofing, which means printing a sample page. If necessary limited corrections, for example increasing the depth of cells, are made by hand in the revision department.

The cylinder is now mounted in the printing machine. Sheet-fed plate machines handle single sheets of paper and are used mainly for high quality colour reproduction, particularly of fine art work. Reel or web-fed presses require a continuous band of paper and are used for long run colour work to high standards of quality. They are predominantly used in the magazine and colour catalogue field, for postage stamps and for printing on non-absorbent packaging materials such as film and aluminium foil.

On sheet-fed presses the ink is transferred from the ink reservoir by a fountain roller which floods the plate with ink, while on reel or web-fed processes the cylinder may revolve in an ink trough. In both cases the excess ink is removed from the metal surface by a steel doctor blade. Because the ink must flow in and out of the minute cells, gravure inks are thinner than those used in letterpress and lithographic printing and consist of finely dispersed mixtures of pigment, resin and solvent. After transfer to the printing paper or other material, the solvent evaporates, leaving a solid

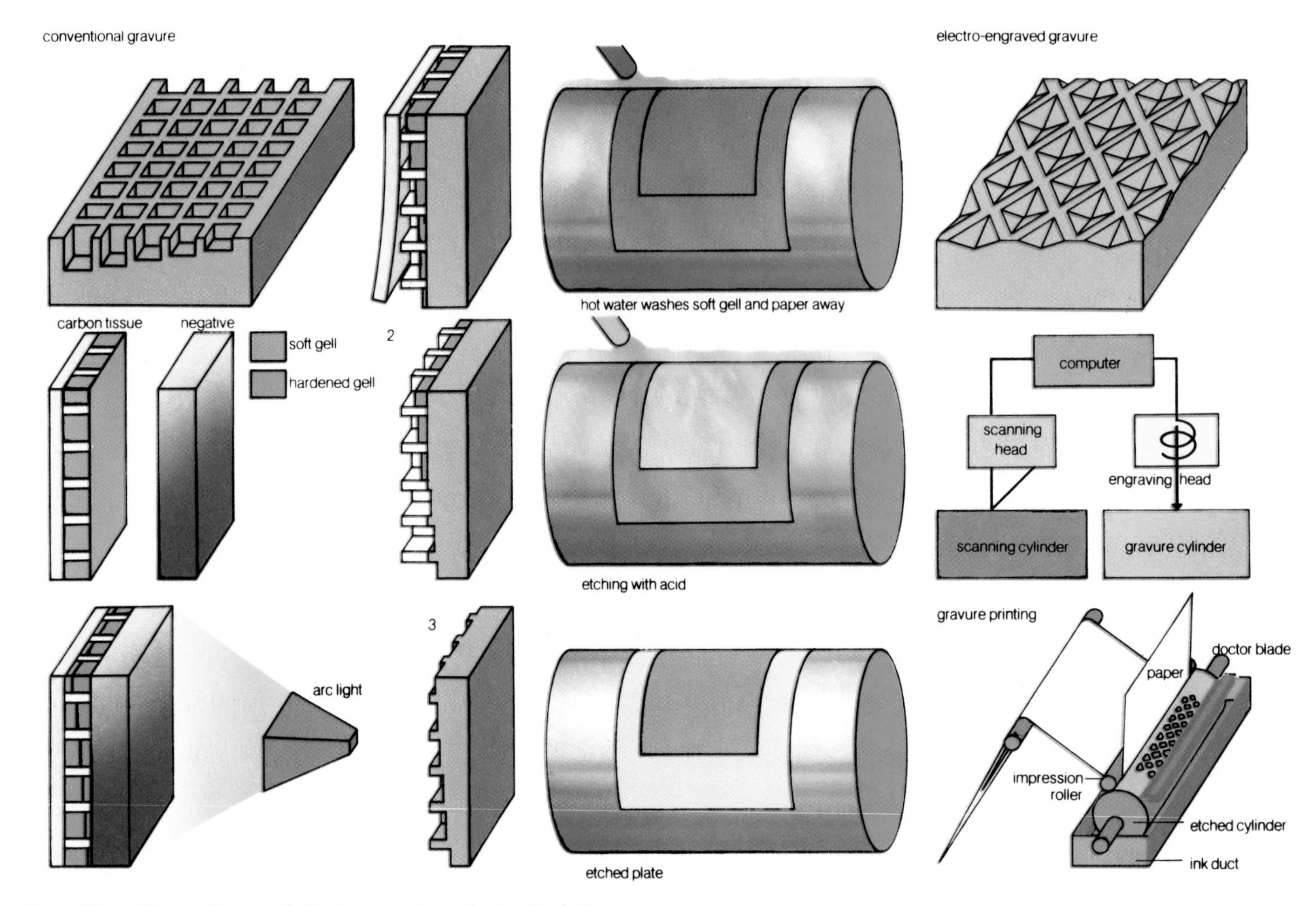

ink film. Some form of drying system is included on most gravure presses to accelerate the evaporation of the solvent. This is particularly important with high speed work. The ink is circulated by pumps to prevent the formation of a skin and to counteract the tendency of some pigments to settle out. A large volume of ink is circulated to reduce the need for constant adjustment of the viscosity by the addition of solvent.

In web-fed presses the printing paper passes at high speed between the copper cylinder and a rubber roller to which pressure is applied by an impression cylinder. A large rotary gravure press comprises a number of adjacent units each printing one colour, so for four colour printing (normal colour printing is done in cyan, magenta, yellow and black) on each side of the web eight units would be needed. At the end of the press is a large unit where the web is cut, folded, collated and stitched or stapled to form the completed magazine. Nowadays there are rotary gravure presses capable of producing 55,000 copies per hour of a 48 page, four colour magazine.

LITHOGRAPHY Lithography is a printing process which is based on the fact that grease and water do not readily mix. The image areas on a lithographic plate do not stand up in relief as in the letterpress process, and neither are they recessed into the plate as in gravure. The plate simply relies on the fact that the image areas attract a greasy ink while the non-image areas repel it. In order to prevent any ink adhering to

the non-image, these areas are coated with gum and kept slightly damp, so another of the distinctive features of lithographic printing is that it involves the use of water on the press as well as ink. Before each impression is made a litho plate must be dampened and then inked. First, rollers in the damping system distribute water from a water trough (called the fountain) to the non-image areas of the plate, then inking rollers transfer a film of ink from the ink reservoir (also called the duct) via a series of distribution rollers to the image areas of the plate. The balance between the water and the ink applied in lithography is a critical factor which calls for a great deal of craftsmanship.

Although lithographic printing was discovered in about 1800 by Alois Senefelder, it is only in recent years that the process has acquired such commercial importance. It has now overtaken letterpress as the major printing process and its applications range from the small offset duplicating machines in offices to the large web-offset installations producing magazines and newspapers in colour.

In the original lithographic process, the printing area was made by drawing directly on to a polished limestone surface with a greasy crayon, and these origins explain the derivation of the word lithography from the two Greek words, *lithos* meaning a stone, and *graphe* a drawing. Litho stones were incorporated into flatbed presses but today's fast litho presses all use the rotary principle. This means that litho printing surfaces must be made on thin, fairly strong, sheet materials which can be curved around the cylinder of a rotary press. A great variety of materials are used including aluminium, zinc, plastics, paper, and combinations of metals such as chromium and copper. The most important of these plate materials is undoubtedly aluminium, which combines lightness, flexibility and reasonable strength with excellent lithographic properties and a relatively low cost.

In their simplest form, litho plates can be made by drawing directly on to the metal surface with a greasy crayon, a method which is still used for limited editions of artists' lithographs. Another example of a direct image plate is the short run plate for an office duplicator made by typing on to a paper plate using a special total-transfer ribbon. In both these cases, however, the litho image is very susceptible to wear and the plate has a comparatively short press life. The real breakthrough for lithography came with its marriage to photography and today, with few exceptions, litho plates used in the printing industry are produced by photo-mechanical means.

The majority of litho plates are made of aluminium. The surface of the metal is normally roughened or grained by mechanical or chemical means in order to increase its surface area and so improve its water-holding capacity in the non-image area. This is taken a stage further in the use of anodized aluminium, on

Top left: in conventional gravure, a positive is photographed on to screened carbon and light hardens the gel to different extents. The image is etched through pores in the gel. In electro-engraved gravure, the image is monitored on a scanning cylinder, the signal controlling the engraving head on the gravure cylinder.

Centre left: lithography. A letter inked with a water repellent ink surrounded by a film of water on the printing plate.

Bottom left: a negative working surface litho plate is made by mounting negatives on a plastic film, which is cut and peeled away in the image areas. This is placed with the plate in a vacuum frame (shown here) and exposed, using a carbon arc.

Right: of these 16 rollers in a small offset flatbed printing press, only the lowermost four actually touch the plate; the rest are for spreading the ink.

which the aluminium oxide layer has been built up electrolytically to provide an extremely hard yet porous surface, which assists clean litho working and a long press life.

The first step in litho platemaking is to apply a light-sensitive coating to the surface of the metal. This is done by pouring the coating solution on to a plate while it is being spun round on a revolving table known as a whirler. The spinning action of the whirler ensures that all parts of the plate are coated to a uniform thickness. 'Wipe-on' plates and solutions are also available where the coating is applied by hand with a sponge. Since coating, however, is a specialized process, the trend in recent years has been for the platemaker to buy plates that have already been coated with a suitable light-sensitive material. These are known as pre-sensitized plates.

All photomechanical methods of making litho plates depend on the fact that when a coated plate is exposed to a powerful light source through a photographic negative or positive, the areas of the coating struck by light are changed chemically. Depending on the type of light-sensitive coating used, they will either become more soluble or less soluble in a particular solvent After exposure the more soluble areas can be washed away to leave a hard stencil of the image on the plate. Plates are known as negative-working or positive-working, depending on whether exposure is made through a negative or a positive.

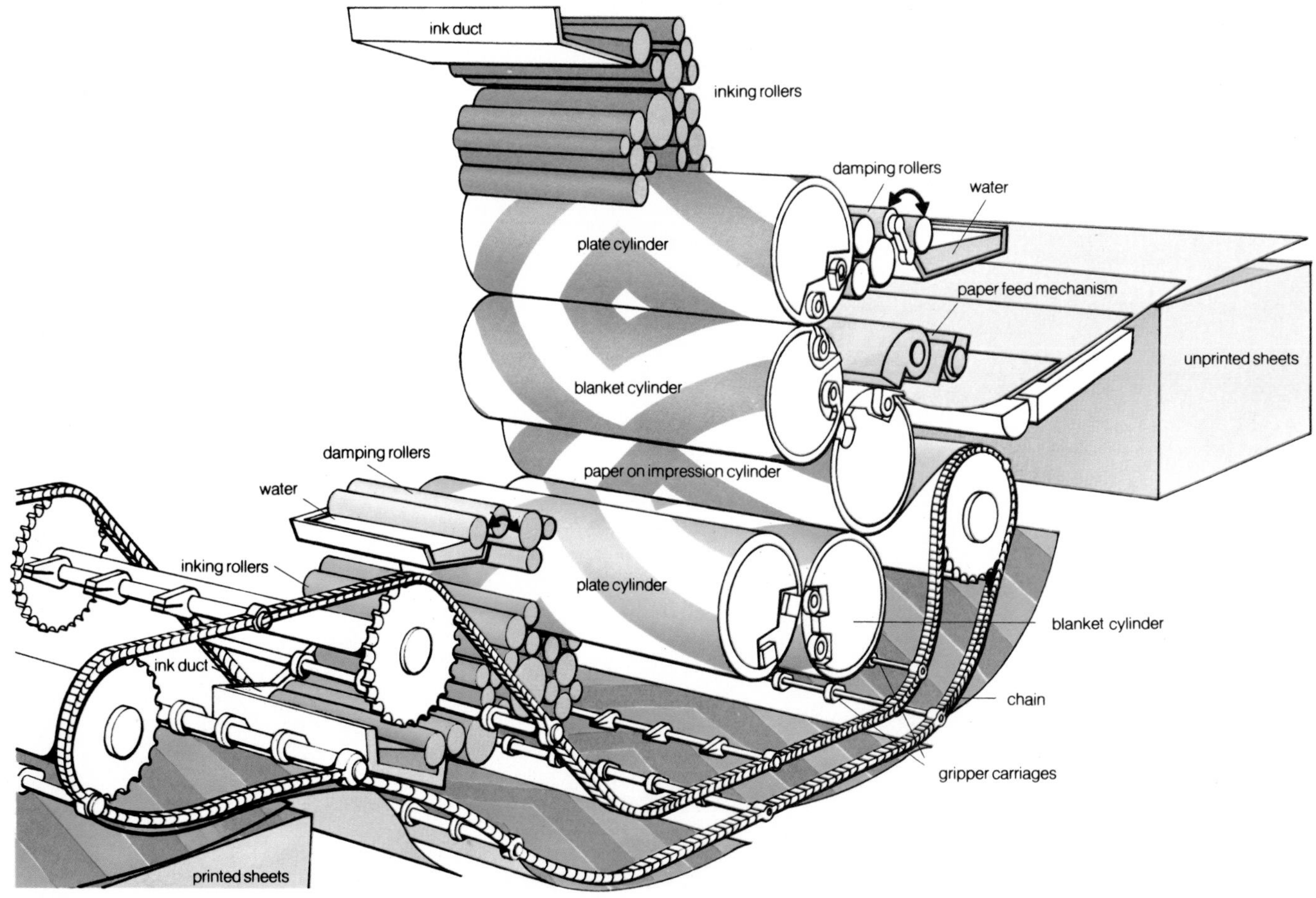

The original subject to be reproduced must be converted into a line or halftone negative or positive. The type of photographic film suitable for litho platemaking has a high degree of resolution together with extreme contrast and is known as lith material. This produces negatives or positives on which areas of solid dense black contrast sharply with areas of clear film. Halftone dots with solid centres surrounded by fringes due to graduations of density are not suitable for litho platemaking.

The light sensitive coatings used in platemaking have traditionally been natural colloidal materials such as gum arabic or egg albumen, sensitized with ammonium dichromate. In recent years these natural substances have tended to be replaced by new synthetic materials like polyvinyl alcohol, diazo compounds and photo-sensitive synthetic resins. Litho plates fall into three broad classes; surface, deep-etch, and bi-metal.

In making a negative-working surface plate the coated plate is placed in a vacuum printing frame and the negative positioned accurately over it. The vacuum then pulls the negative into close contact with the plate during exposure to a powerful light source. Light passing through the clear areas of the film hardens the coating. The unhardened areas are then washed away, leaving the hardened areas to form the image of the plate. These image areas stand slightly proud of the surface and they are subject to abrasion during printing. Surface plates have been used for short-run work in the past but pre-sensitized surface plates have been developed which are capable of runs of 50,000 and more.

In producing deep-etch litho plates, the coating is exposed to light through a positive. The unhardened areas are washed away, but in this case the function of the hardened areas is not to act as the litho plate but to form a protective stencil while the image areas are lightly etched and then filled with a hard lacquer. Despite the name deep-etch, the image areas are only about 0.008 mm below the general level of the plate. The fact that the image is slightly recessed, however, makes it more resistant to wear and hence better able to stand up to press runs, which can exceed 100,000.

Bi-metal plates take advantage of the fact that oil and water do not wet all metals with the same ease. An

Top left: washing off dust during a printing run.

Bottom left and right: diagram and photograph show the same machine, the Roland Parva, a sheet-fed press for two-colour printing which can handle 10,000 sheets an hour. In the diagram, some inking and damping rollers are drawn cut off halfway for clarity. After feeding to the first cylinder, each sheet of paper is drawn through the mechanism by grippers on a pair of endless chains. The damping rollers move between the water rollers and the others; this movement is controlled to adjust the amount of water transferred.

oleophilic (oil-loving) metal, such as copper, is used for the image and a hydrophilic (water-loving) metal, for example chromium, for the non-image areas. The method of platemaking is similar to that used in the deep-etch process but the purpose of etching is to remove a thin layer of chromium to leave bare copper exposed to form the image. Although bi-metallic plates are more expensive than those described earlier, they are far more resistant to abrasion and they are capable of press runs exceeding a million impressions.

All modern lithographic printing is offset printing, that is, the plate does not transfer ink directly to the paper but first offsets the image on to a rubber 'blanket' from which it is printed on to the paper or some other substrate. The offset principle was first used for the printing of tin plate in order to overcome the problems met in printing directly from metal to metal, but it is now used for all substrates. Indeed, the word offset has become synonymous with lithography, and this is unfortunate because the offset principle is also used in letterpress and even in gravure printing. The use of an offset blanket reduces the wear on a litho plate and its resilience makes it possible to print fine halftone images on a wide variety of papers, including those with fairly rough surfaces. Another advantage of printing offset is that the image on the plate can be the right way round and not 'wrong-reading' as in any direct printing process. Offset blankets consist of several layers of woven fabric laminated with thin layers of synthetic rubber. They are made in various thicknesses and they can be designed to provide various combinations of hardness, resilience and resistance to ink solvents and oils.

Although flatbed offset presses are generally used for proofing purposes, all production machines utilize the rotary principle. The basic printing unit of a sheet-fed offset litho press consists of a plate cylinder, a blanket cylinder and an impression cylinder. The plate is clamped round the plate cylinder, and with each revolution of the machine it first comes in contact with a damping roll and then with the inkers. The paper passes between the blanket and the impression cylinder but it does not touch the plate. Sheetfed offset presses are commonly available with two, four or six printing units for colour work. Their running speed depends on the type of machine, the nature of the work being printed and the paper being used, but it can be as high as 10,000 sheets per hour.

Web offset presses print on a continuous web of paper, which is fed into the machine from a reel. Each printing unit has the same basic elements as those on sheet-fed presses, but there are many ways in which units can be combined to print one or more colours, possibly on both sides of the paper. Web offset machines have proved particularly successful in printing provincial and some national newspapers in colour.

PRINTING PICTURES The earliest printed illustrations were from woodcuts, which actually preceded movable type. Later woodcuts and metal type were used together in letterpress printing. In making

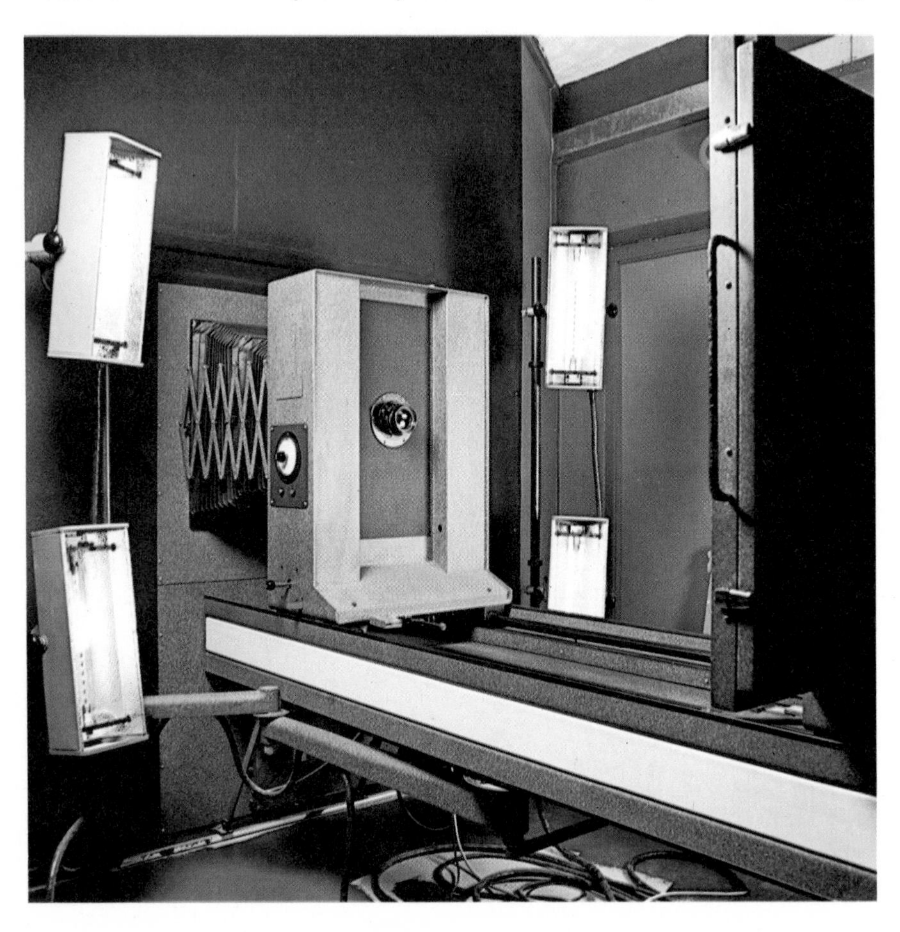

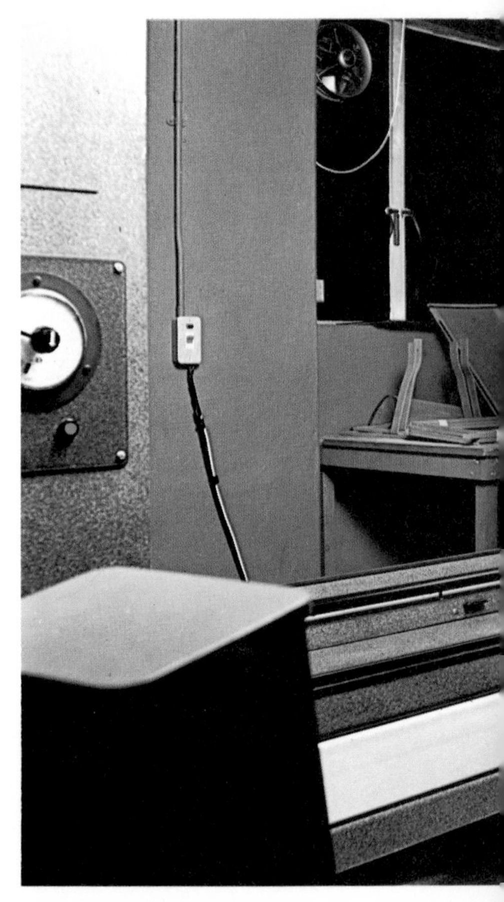

woodcuts, designs were drawn in reverse on the surface of the wood and the intervening spaces cut away. Early woodcuts were of simple design but artists like Dürer and Holbein exploited the method to great effect in the reproduction of their drawings. At the end of the sixteenth century woodcuts were largely superseded by copper engravings but much later the art was to be further developed particularly by Bewick, whose fine wood engravings allowed the reproduction of tonal detail, so anticipating halftone photography.

The letterpress and lithographic processes are unable to print different ink thicknesses and hence different densities at different points on a sheet of paper, so the effect of continuous tone has to be created as an illusion by breaking up a picture into a system of small dots, the size of which is varied depending on the tone required. A very light or 'highlight' tone is reproduced by printing very small dots which will be separated by a large amount of white paper. A very dark or 'shadow' tone is achieved by printing large dots which will almost join up into a solid area of print. This method of creating the illusion of continuous tone is known as the half tone principle. It relies on the fact that the human eye will not distinguish each dot but rather sees a mixture of the total amount of light reflected from the ink and surrounding paper in a given area. When, as in newspaper printing, a coarse half tone screen is used with say only 65 dots to the inch, the eye is just able to distinguish individual dots. At the other extreme, fine halftones with as many as 250 dots to the inch are sometimes used in the high quality reproduction of illustrations. The dots are printed in diagonal, lines since these are less obvious than vertical ones.

The conversion of continuous tone subjects into a halftone form may be carried out in the camera with a crossline screen or more usually with a contact screen. The contact screen consists of a regular pattern of grey or magenta dots on transparent film, the edges of the dots fading away (vignetted) rather than being sharp. The screen breaks up the light reflected from an original into spots of light of varying size and brightness. The darker areas of the original reflect little light so small dots are produced; the lighter areas reflect more light and so produce larger dots. This depends on the use of lith film in the camera, which has very high contrast and therefore turns the vignetted dots into sharp ones of various sizes.

The crossline screen has a grid of sharp bars. This is held at a small distance from the film, thus producing the vignetting effect, whereas the contact screen has to be in close contact with it, as its name implies. This is usually accomplished by a vacuum frame, which holds the films closely to it by suction.

Below: left to right: a series showing a process camera used to photograph originals at the right size for printing. At the left is bellows and lens on a track, pointing towards the original picture (centre). In the next room (right) the image is sized up on a ground glass screen before the film holder is put in place.

Most colour illustrations are reproduced by successive printings of yellow, magenta, cyan (a greenish blue) and black inks. By combining these colours in suitable proportions it is possible for most colours to be reproduced. A different printing plate must be produced for each colour. This is done by photographing the original through appropriate coloured filters to give separation negatives which are then used to make the four printing plates. The individual screens all have to be at equal angles to one another to avoid patterns appearing, and the plates all have to be printed in very accurate register so as to preserve the even dot structure and maintain sharp edges.

PRINT IDENTIFICATION The identification of the printing process used to produce a given piece of work can be very difficult, but there are a number of points to look for. The task is easier with the use of a good magnifying glass.

A letterpress print may show a slightly embossed effect on the reverse of the sheet; ink squash may have caused the thickening of the ink film round the edges of a letter or a halftone dot; and the general effect is of crisp, bright and sharp detail. Litho prints tend to be less contrasty, halftone dots have fuzzy rather than sharp edges, there is no ink squash or embossing effect and very small highlight dots can be seen even on fairly rough paper. Prints produced by the conventional gravure process show a pattern of square cells of equal surface area but varying ink film thickness; type matter may have a saw tooth edge and solid areas may show a pattern of white spots or lines where ink has failed to flow sufficiently to eliminate the cell pattern.

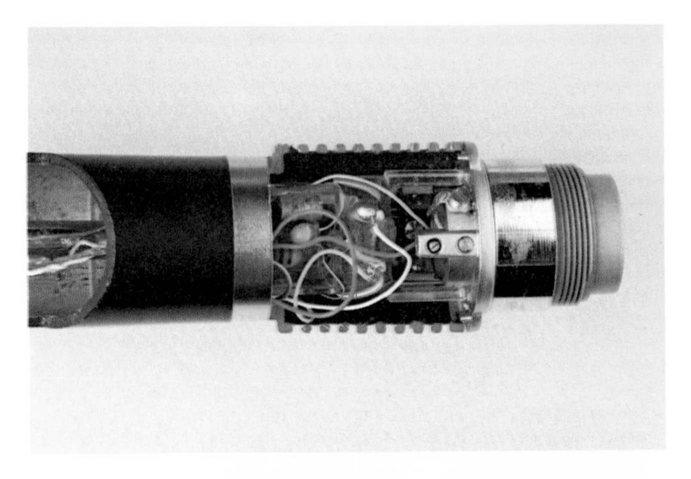

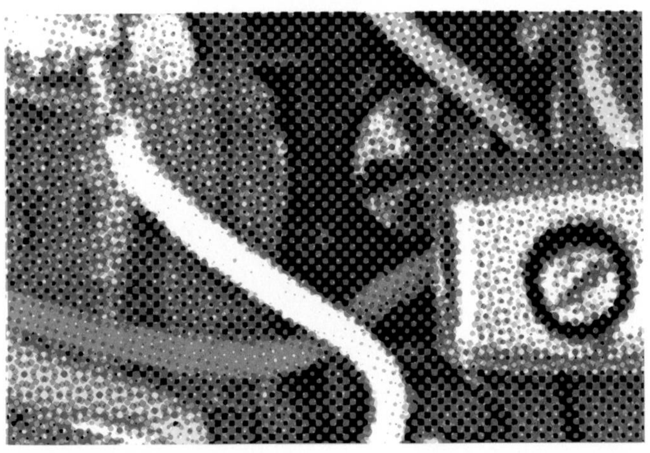

SELECTING THE PRINTING PROCESS In letterpress printing, ink is transferred from a raised surface, in gravure the ink flows out from small cells in the printing plate and in lithography printing takes place from a flat surface on which image areas attract ink and non image areas repel it. Printing presses range in size from the small offset machines used in many offices to the huge web-fed installations producing newspapers and magazines. Letterpress machines may be based on platen, flatbed cylinder or the rotary principle. Gravure and litho presses all use the rotary principle which allows higher printing speeds. In selecting the most suitable method of production and type of press for a particular printing job many factors have to be taken into account. To start with one must ask the right questions about the job itself. What quality is required? What is the length of run? Are illustrations involved? How many colours are to be printed? What quality of paper or other base is to be used?

Letterpress and offset lithography are the two most versatile processes and in each case the quality of reproduction can reach a high standard. The fact that ink is applied to the base by a soft and resilient rubber covered cylinder means that a wide variety of papers, boards and even tinplate can be printed by offset litho. This is not the case in letterpress, which demands the smooth surface of a coated paper for the printing of fine screen halftones. Machine speeds are normally higher in litho than in letterpress printing, although machine stoppages tend to be more frequent and more waste sheets are printed. The litho process is particularly suitable for jobs involving a large number of illustrations, since the litho printing plate is always made via a photographic negative or positive; on the other hand, preparatory costs in letterpress are relatively low compared with the other processes, especially if no pictures are involved. As type and blocks are movable within a forme, the letterpress process allows changes to be made at the last minute or even during the course of a printing run, but unfortunately type metal is heavy, bulky and expensive to store. This problem is being overcome by the trend towards the assembly of letters on photographic film, although this means the loss of flexibility in making corrections.

Photogravure is an excellent process for the reproduction of illustrations in monochrome or colour, giving good depth of tone in shadow areas and a

Left: the upper picture is reproduced normally in four colours with a screen of 5.9 dots per mm (150 per in.). Below it is an enlarged view of a detail: the separate magenta, cyan, yellow and black dots can be seen. Each colour screen is at a different angle by 15° to avoid a moire (mottled) effect; the black screen is at 45°, but it is much lighter than the others, simply adding contrast and weight to the other colours combined.

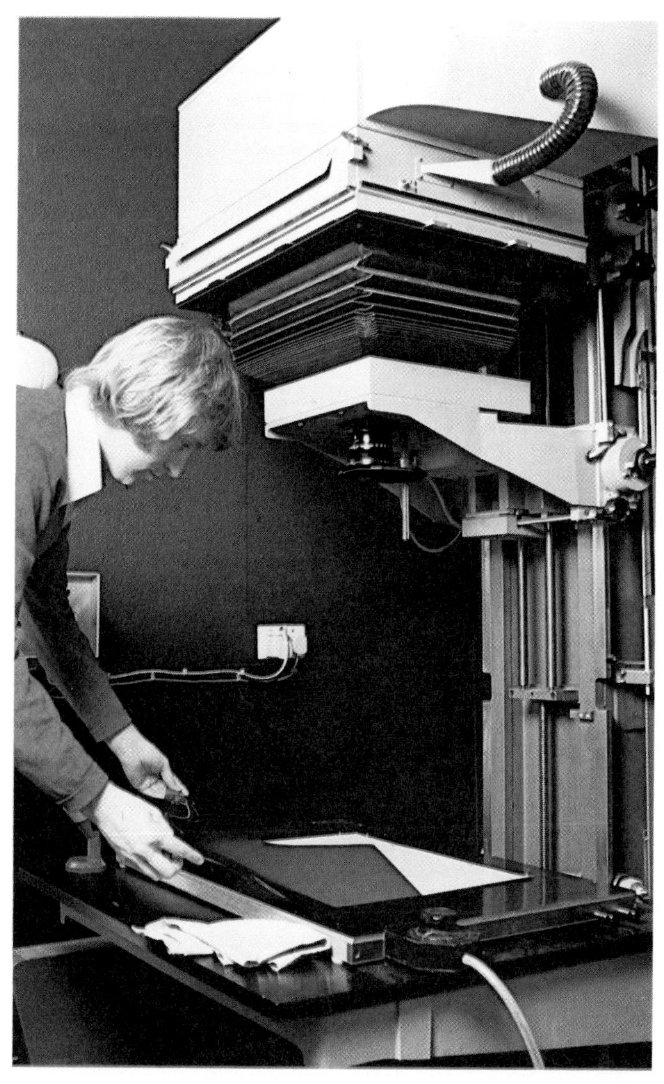

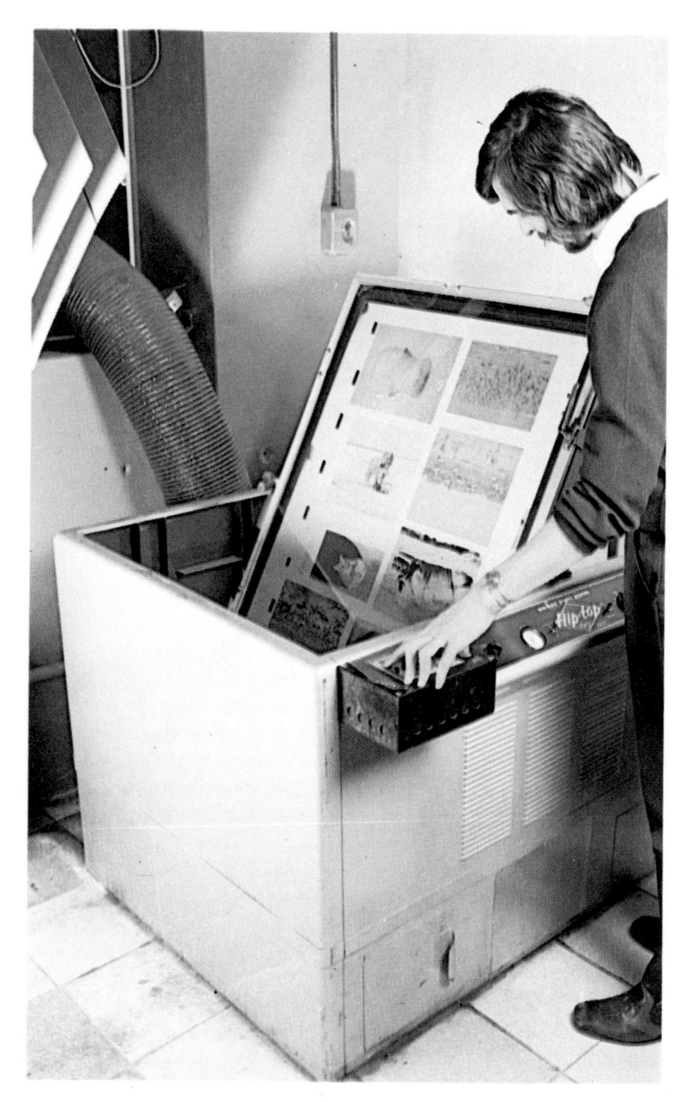

Above left: to make a negative from a transparency, an enlarger is used. The transparency may be any size up to 200 x 250mm (8 x 10in.) and is placed in the carrier just above the operator's head. The image is projected on to the baseboard, where the operator is placing a sheet of film underneath a contact screen. This is normally done in red light, to which the film is insensitive.

Above right: from the negatives, positives are made by contact printing. These are then exposed on to a plate by ultra-violet light.

delicacy and softness in the middle and lighter tones. The costs of plate and cylinder making are very high, however, so the process is best suited to very long runs on web-fed rotary presses; hence its use in the printing of large circulation colour magazines, catalogues, postage stamps, and books such as the one you are reading.

NEWSPAPER PRODUCTION Newspapers began as pamphlets and broadsheets in the 17th century. Their original function was the dissemination of news and opinion, but advertisers quickly found them useful. Technical developments in the Industrial Revolution and the growth of the popular press in the 19th century have resulted in the production of newspapers as we know them today, using techniques which had not changed for many years until the present development of electronic newspaper production.

Newspaper advertising is in two categories: classified and display. Classified adverts appear under special categories and are usually printed in a self-contained section; display adverts are intermixed throughout the paper with editorial content. The number of pages in a newspaper is based on a predetermined ratio of advertising to editorial matter.

Above left: a proof is made of a newspaper page on a flat proofing press. Any necessary corrections or alterations are then carried out.

Above right: a curved printing plate is made from the original flat plate by taking a moulding from it in a flexible papier-mache like material called flong, and bending this to shape before casting the plate against it.

Since the advertising layout can be determined well in advance of the news, a plan of each page showing the advertising layout is given to the editorial department to assist in the layout of the paper.

Classified advertising is sold over the telephone; display advertising is obtained by salesmen in the field, supported by an art department which can prepare artwork as necessary. Advertising is indispensable in modern newspapers because the production costs of the paper could never be met by newspaper circulation alone.

The gathering, selection and editing of the news, as well as the writing of leaders [editorials] carrying the commentary, is the function of the editorial department. Some newspapers keep the straight reporting of the news separate from the editorial comment, and some newspapers allow the news stories to be written according to the editorial opinion. Sometimes the story will carry a by-line (the reporter's name).

The Editor usually has a conference each day with his aides to determine the relative importance of the news coming in; a rough plan is made and kept under review as the news breaks. The news comes in from a variety of sources, including local reporters in the field, often working on special assignments such as sports or financial news, and international news agency wire services. Local news often goes directly to the news desk, while international news goes to sub-editors.

The British sub-editor is called the desk man or copy reader in the USA. His job is to help select material according to its importance, since far more news comes into the news room each day than can be printed in the paper. The British sub-editor often rewrites a story, condensing it to save space; his north American counterpart will more likely write in background information, so that the reader seeing a developing story can catch up on it. An American news story is often written in such a way that the last few paragraphs are short and not so essential, so that they can be cut if necessary for space reasons.

Apart from the mainstream of the news, feature articles on a wide range of subjects, often suggested by the news, are prepared by writers in the feature department. The taking and selection of pictures is organized by the picture editor.

Copy (written news) is typed in the office or received over the telephone by copytakers who type stories dictated from all over the country or even abroad by reporters and correspondents. In the wire room, or telegraph room, a nonstop flow of news and pictures comes in through a worldwide communications network. Teletype machines are operated by keyboards at the other end of the wire, using radio links and submarine cables between the continents. Pictures are sent out by fitting a photograph to the drum of a transmitting machine. The drum rotates and the shades of black and white in the photograph are converted into electronic impulses of varying intensity which are converted at the receiving end back into a photo by a similar drum holding photo-sensitive paper.

The major wire services are Reuters, Associated Press and United Press International. In addition, there are national counterparts such as the Press Association serving Great Britain, and Extel, which supplies sporting and financial news. Many of the largest newspapers also operate their own press service, sending their news and features to subscribing papers.

Copy is sent to the composing room, where it is set into type. Before 1886, each letter was set by hand in lead. Since then the bulk of lines in the newspaper have been set by Linotype. (Headlines are usually set on a Ludlow machine, using the same method except that the matrices are assembled by hand.)

Since World War II, Linotypes have been adapted to be operated by punched tape. Since the late 1950s computers have been used to justify lines. Computerized typesetting, called teletypesetting or TTS, has tripled the capacity of the Linotype machine so that it can produce ten or fifteen lines a minute.

When the copy is set, a proof sheet is printed and sent to readers so that corrections can be made. The corrected setting is then sent to the stone, or page make-up area, where it is fitted into page forms. Photographs are fitted into the page in the form of zinc half-tone blocks, prepared by photo-engraving. The picture is re-photographed through a prism, so that the new negative is a mirror image of the picture, and through a screen which is divided into fine lines. The screen breaks up the light into dots of varying intensity according to the light and dark parts of the photo. (Cartoons and line drawings are exposed without the screen, producing a line block). The negative is placed over a zinc plate which is coated with photo-sensitive chemicals. Brilliant light from arc lamps bakes the coating on where it passes through the negative; elsewhere it remains soluble. The soluble parts are washed away and the plate is bitten in a bath of acid.

The assembled page is taken to the stereotyping department, where printing plates are made from it. A sheet of papier-mâché-like material called a flong is placed over it and subjected to pressure in a moulding press. The resulting mould, also called a matrix, is curved, dried and placed in a casting machine, which uses it to make a semicylindrical metal plate which weighs about 40 lb (18 kg) and is a mirror image of the newspaper page. This curved plate is mounted on the letterpress which prints the paper. If a newspaper has a large circulation, it may have several presses and several plates will be required; the plate-casting machine has been automated over the years so that it can produce up to seven plates every two minutes.

The rotary press may be up to 100 feet (about 30 m) long and weigh 300 tons. Using reels of newsprint weighing a ton and carrying five miles (more than 8 km) of paper, it can print up to 70,000 copies of the paper per hour. The press also cuts and folds the newspapers, and automatic equipment is used to tie

Below: the first successful rotary machine was the Hoe Type-Revolving Printing Machine, invented in 1847. This one, imported from New York in 1860 in 47 crates, and weighing 30 tons, could print 25,000 copies an hour.

Centre: electronic editing and proofing terminal.

Bottom: pressroom at Yorkshire Post Newspapers Ltd.

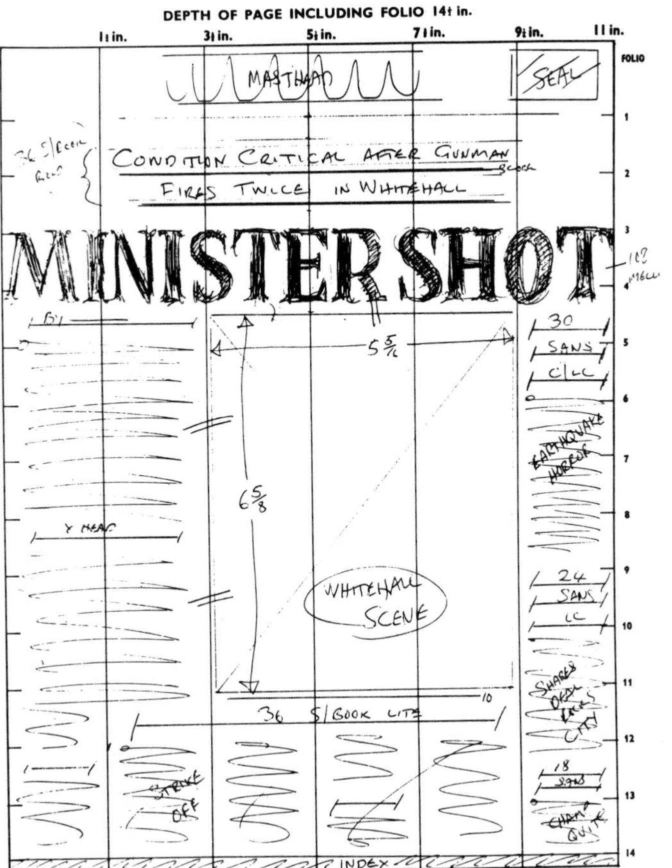

them in bundles or put them in plastic bags. Most newspapers distribute their own product, but some larger papers use wholesale distributors who send the papers all over the world.

The letterpress printing technique described above has been used for over half a century without fundamental change; most of the newspapers in the world are still produced that way. Using electronics and computers, changes are now taking place which allow faster production, savings in manpower and other costs, and better reproduction of pictures, especially colour. The key to these changes is the switch to offset printing. In the USA, where developments have been more rapid, more than half of the 1761 daily newspapers have switched to offset printing; the largest newspapers are still printed using the letterpress method because of their investment in presses, which may last for fifty years.

For the same price as a Linotype machine, the photocomposition machine can set lines of type ten times as fast. The make-up operator assembles a paste-up newspaper page, using columns of type produced by the photocomposing machine. (The machine can also mix sizes and styles of type, unlike the Linotype.) The paste-up is then photographed, and the negative is exposed to the offset plate, an aluminium plate coated with photo-sensitive material. This litho plate is not a relief plate but a flat or planographic one; the press applies ink and water to the plate; the image area of the plate accepts the ink but

Daily Mail

3p

WIN A CAR FORTNIGHT Page 12

Condition critical after gunman fires twice in Whitehall

MINISTER SHOT

Daily Mail Reporter

A GOVERNMENT Minister was shot down by a gunman in Whitehall last night.

As two policemen disarmed the attacker a police car rushed the unconscious Minister across Westminster Bridge to St Thomas's Hospital, where his condition was later said to be critical.

Witnesses said two shots were fired by a fair-haired young man dressed in a black shirt and jeans. He was dragged, struggling, to Cannon Row police station and questioned by Special Branch men. He is thought to be an American.

The news was telephoned to the Queen at Balmoral. As a shocked House of Commons was told of the shooting by the Speaker the hospital issued its first bulletin:

Bodyguard

Arrests

Police hold back crowds in Whitehall after last night's shooting.

Earthquake hits five more cities

OVER 5,000 people are feared dead in the earthquake which devastated large areas of Northern Chile yesterday.

Share deals halted by City probe

Engineers' strike called off

Admitted

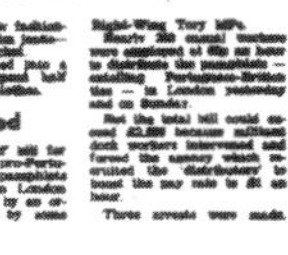

I'll quit, says new champion

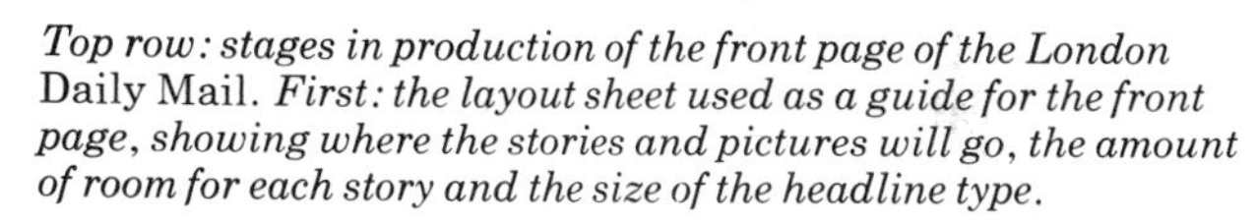

Top row: stages in production of the front page of the London Daily Mail. *First: the layout sheet used as a guide for the front page, showing where the stories and pictures will go, the amount of room for each story and the size of the headline type.*

Second: the assembled page forme. It is from this that the proof is taken, as shown on p. 152.

Third: the flong, or matrix, being removed from the page forme. Pressure in a moulding press causes the soft material to take the impression of the forme, but the surface coating is strong enough to retain its shape when the flong is curved to make the curved printing plate.

Fourth: the completed front page as actually printed.

Below, far left: the semicylindrical metal printing plate, which is a mirror image of the completed newspaper page, being removed from the casting machine. It weighs about 18kg (40lb.).

Near left: installing the reels of paper on the press. This is low-quality newsprint paper made of mechanical wood pulp, since it is not required to last. Each reel is about 8km (5 miles) long and weighs about a ton.

Above and below: two views of the press at the Washington (DC) Evening Star. A letterpress machine, once installed, may last fifty years. In cities with more than one newspaper, they sometimes cut costs by sharing the printing plant and advertising revenue, while remaining independent editorially.

rejects the water; and the image is transferred on to a rubber blanket which then transfers it to the paper. Printing plates have been developed which are produced photographically but can be used to make a flong, or mounted directly on a letterpress, making possible a combination of the methods.

The development of offset printing has encouraged the greater use of electronics and computers. Photocomposition machines are built which can set 500 lines a minute. The news can be stored on magnetic discs and shown on a video display terminal (VDT) which uses a cathode ray tube similar to a television screen. The copy can be reviewed, corrected, and brought up to date while it is still on the computer. When the copy is ready it can be sent by wire or punched tape directly to the photocomposing machine. Other systems use optical character recognition (OCR), in which the typing is done on a special machine whose output can be read by the photocomposer. The new machinery eliminates the need for proofs, retyped corrections and hot metal process, as well as storing copy in the computer until it is needed, resulting in enormous savings of time and manpower.

Photocomposing machines are being developed which will reproduce entire pages, with photographs and art work, eliminating much of the paste-up work. Laser beams will be used to engrave plates; these will

be operated directly from the composing machine, eliminating the need for the camera which now produces the page negative. The printing press itself may some day be replaced by an electrostatic process similar to Xerox copying.

In the future, the problem of rising cost and scarcity of newsprint, together with distribution of newspapers where transport is difficult, will also be a focus of research and development along with production itself.

RECORDS

In recording, sound waves are transduced into electrical energy by the microphone, amplified and recorded on magnetic tape over the frequency range 30 to 20,000 Hz. Various electrical compensations have to be made in order to accommodate the frequency range in the groove of the finished record and to reduce tape hiss; these compensations are reversed in the playback amplifier, and they must conform to an internationally agreed standard so that the finished recording will sound good on any reproducer.

Modern stereo-recording techniques involve the use of several microphones recording up to 24 channels of information at 15 inches (38 cm) per second on a two inch (50 mm) wide magnetic tape. Mixing is the process of blending the channels to result in the desired musical balance, and transferring them to a ¼ inch (6.2 mm) two-track tape which is called the master tape. Recordings are taken in short sections and the master tape results from an editing process, in which the artists, the producer and the engineers all take part.

MASTER DISC The beginning of the process of manufacturing records is the production of a master lacquer disc. The recorded magnetic data is transduced back into electrical signals which are applied to an electromechanical device holding the groove engraving tool (a sapphire or ruby cutting stylus). The precision cutting lathe incorporates a heavy turntable rotating at constant speed; the plastic medium into which the groove is cut is a nitrocellulose lacquer applied to an aluminium disc substrate.

An adjustable micro weight device ensures that the groove is cut to a constant depth, while the cutting stylus is simultaneously modulated vertically and laterally, producing the 45/45 stereo groove which was patented in Britain in 1931. The cutting lathe is highly automatic and will, for example, regulate the spacing of the spiral groove according to the level of modulation. This technique is called variable pitch, and makes possible a playing time of up to 35 minutes on one side of a 12 inch (30.5 cm) 33⅓ rpm record. The cutter moves inwards on a lead screw as the disc rotates, and the pitch (number of grooves per linear inch) varies from 140 to 400. The plastic thread, or cutting swarf, is removed continuously by suction applied close to the cutting point. A smooth surface on the walls of the groove is ensured by the quality of the lacquer for-

Above: the black lacquer master disc is cleaned and treated chemically to make a spray of silver nitrate solution adhere; after spraying it is placed in a tank, shown here, to be silver plated.

Below: the silver plated lacquer disc being removed.

Left: a shell about to be placed in a solution for electroplating. When the lid is closed, the disc is spun in the solution, which is constantly filtered and monitored for purity. Cleanliness is vital.

Above: the positive shell being split from the stamper. The stamper will have its back polished smooth, an optically centred hole punched in it, and be machine pressed to give it a raised edge and label area.

mulation, the geometry of the cutting stylus, and the heating of the stylus.

The cut groove dimensions are a mean depth of 0.002 inch (0.05 mm), and an included angle of 90° with a bottom radius of 0.0002 inch (0.005 mm). The reproducing stylus dimensions are an included angle of 59° and tip radius 0.0006 inch (0.015 mm). (The term 'needle' is incorrect for the playback stylus because it is not 'sharp'; the tip is ball-shaped and the stylus should ride in the groove with each side touching a groove wall, rather than in the bottom of the trough.)

MASTER SHELL The lacquer disc is chemically cleaned and the surface is rendered hydrophilic (water attracting) by treatment with specific surfactants. It is then sensitized with a tin chloride solution. Stannous (tin) ions are absorbed on the surface; these act as nuclei for the silver which is formed by the simultaneous spraying of an ammoniacal silver nitrate solution and a reducing solution.

The mirror-like silver deposit is extremely thin, but is sufficiently conductive to allow a nickel electrodeposit to be formed on it by making it the cathode (negative electrode) in an electrolytic cell containing soluble nickel (nickel sulphamate) and pure nickel metal as the anode (positive electrode). Initially for a short period a low temperature (35°C, 95°F) is used followed by a longer period at 60°C (140°F). The temperature, other chemicals in the solution and the acidity are strictly controlled to obtain a pure nickel electrodeposit with minimum intrinsic stress. After three hours deposition, the master shell is about 0.025 inches (0.625 mm) thick and then is split from the lacquer disc.

POSITIVE SHELL The master shell with its silver face is chemically cleaned and a nickel electrodeposit grown on to it in a manner similar to that described above. In order to be able to separate the two shells, the silver face is first passivated by superficial oxidation or the absorption of an organic colloid. Typically, soluble chromates or natural colloids are used.

The positive shell (also called the mother) can be played, and any small defects that are discovered can be removed using an engraving tool under a stereomicroscope.

MATRIX SHELL The matrix shell is also called the stamper, and is actually used to make the records. It is produced from the positive by the same process as master to positive. The stamper is only 0.010 inch (0.25 mm) thick; its back is polished to remove any small protrusions, it is given a special profile corresponding to the record press mould brick, and an optically centred hole is made in it. (If the hole is off centre the result will be a wavery pitch when the finished record is played; the hole will be in the centre with respect to the edge of the record, but the groove spiral will be off centre, with the result that the tone arm of the record player will sway back and forth with each revolution of the disc.)

RECORD PLASTIC Polystyrene and acrylic polymers have been used to make 45 rpm records (pop singles). The generally accepted polymer for $33\frac{1}{3}$ records has been polyvinylchloride (PVC), usually in the form of its copolymer with about 14% vinyl acetate; this material best fills the requirements of the stylus-microgroove relationship, the physical and mechanical properties of the record, and the conditions normally available for moulding. The PVC contains additives, usually not exceeding one per cent, such as lubricants to ensure perfect release from the mould and to give satisfactory record wear performance, carbon black as a colorant, and a thermal stabilizer to control the decomposition of the PVC. The latter is an organic compound of a metal such as tin, lead or calcium. It is also customary now to add one or more other PVC polymers to increase the melt elasticity, thus improving the mouldability.

The polymer and additives are mixed in a mixing machine operating at 1500 rpm. This results in temperatures of up to 100°C (212°F), so the powder is discharged through a cooling stage to ensure a free flowing 'dry blend'. This is converted into granules, using a twin or single screw extruder which is electrically heated; at the die end, the granules are formed mechanically and cooled in water. The process becoming prevalent now is to feed the dry blend into a small extruder at each record press. The hot extrudate, at about 160°C (312°F), is introduced manually or mechanically into the press.

RECORD MOULDING Semi-automatic presses are still in use but the trend is toward completely automatic record production, which even includes putting the record into a sleeve.

Compression moulding is the favoured process, especially for 12 inch (30.5 cm) long-playing records. The hot 'biscuit' of granular plastic is squeezed, together with the labels, between two die faces which are attached to mould blocks, channelled for heating with steam and cooling with water. The hydraulic pressure developed at the die face is 1 ton inch^2 (140 kg/cm^2) and the steam and water pressures are 140 psi (10 kgm/cm^2). The precise moulding cycle is automatically controlled either electromechanically or electronically, and the moulding cycle takes about 15 seconds for the 7 inch records and 25 seconds for the 12 inch records. The record is removed from the press and the 'flash' (excess plastic) is removed from the edge by rotating the disc against a hot knife.

Many of the 7 inch records are produced by injection moulding; that is, the hot plastic is injected into the mould while the clamping pressure is maintained. It is possible to operate with a double cavity mould, thus making two records in each cycle. Another method is injection followed by compression, the two actions occurring almost simultaneously.

QUALITY CONTROL Quality control in record manufacture is difficult because each record cannot be examined by playing it. Statistical methods are employed in examining samples of a press run, but the real quality control must be practiced in each step of the manufacture.

At the very beginning, the engineer in charge of cutting the lacquer master has to determine the dynamic level at which to cut the disc. If the level is too

Left: pressing a record. The 'biscuit' of extruded PVC and the label are placed in the press. Lowering a window automatically starts the steam-heated, water-cooled machine on its 40-second operation.

Below: this device cuts the 'flash' – excess plastic – from around the edge of the record. The trimmings are melted down to make 45rpm singles.

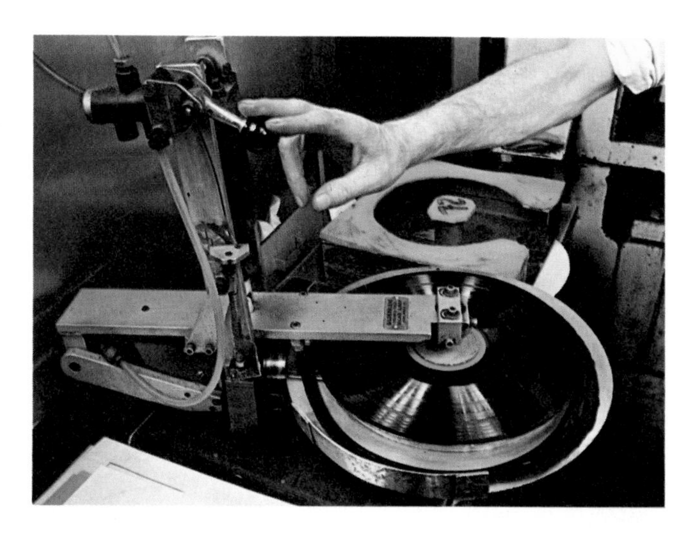

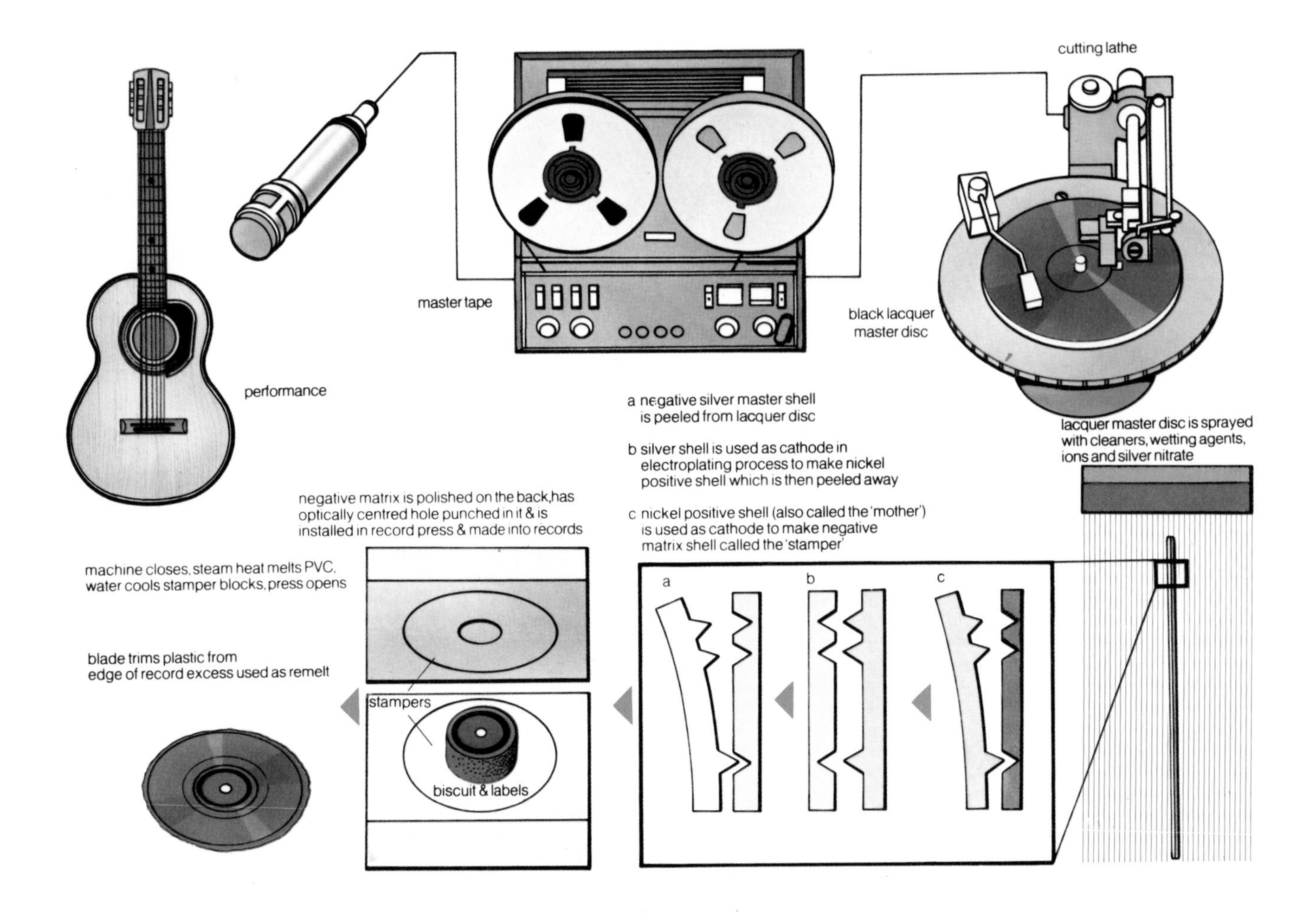

Above: the sequence of events in the manufacture of records. The recording is made initially on a master tape; tape has been used since about 1948. Before then the microphones were connected directly to the disc-cutting lathe. The quality of the finished record depends on the care taken at every step.

high, the grooves will have to be further apart, cutting down playing time, and there may even be distortion and groove-jumping problems in playing the finished record. If the level is too low, the signal-to-noise ratio will be poor. The groove spiral has to have enough land (uncut surface) separating it; if the grooves touch (called *kissing*) groove jumping will result when the record is played; if quiet grooves are too close to highly modulated grooves, pre-echo or post-echo will result (spilling of sound from one groove to another caused by deformation of groove walls). These problems are reduced by the computer which listens to the master tape a second before the cutting stylus and makes the necessary adjustments for variable pitch cutting.

As mentioned above, the various matrixes and shells must be carefully inspected and defects removed each step of the way; above all, they must be clean. The parts can be used many times, but must be inspected and replaced when they begin to wear out; a pair of stampers may be used to make several thousand copies of a record or only a few hundred, depending on the quality desired and many other factors. A record pressed from a worn-out stamper sounds worn-out when played.

In the moulding press, if the back of the stamper is not absolutely clean and smooth when it is installed, the result will be *mould grain*, an imprint on the record surface resulting in low-frequency noise. The timing of the pressing cycle is important. Automation is supposed to help avoid human carelessness, but there is a temptation to speed up the cycle to raise production and save time; if the records are pulled out of the press too soon they will warp as they cool. A common problem in an improper cycle is *non-fill*, or failure of the vinyl to completely fill the groove detail, resulting in cloudy patches of specks on the record surface which make a gritty noise when the record is played.

The recent rise in the price of PVC (along with all petro-chemical products) has forced the record industry to use more re-melt than usual (defective records and returns with the label areas punched out). This makes quality control more important than ever, resulting as it does in small amounts of dirt, paper and other impurities getting into the presses.

ROPEMAKING

Rope may be described as a cord at least 4 mm (0.16 inch) in diameter made by closing (twisting together) three or more strands which are themselves formed of twisted yarns.

Rope was one of man's earliest inventions, and in many respects it has changed very little over the centuries. Materials, however, have varied and nowadays some 90% of ropes are made from synthetic or manmade fibres. The oldest cordage discovered (about 1000 BC) was of flax fibre, although cave paintings of 1800 BC show a twisted rope structure being used for climbing. Excellent examples of rope made from papyrus or Nile reeds may be seen in the British Museum, and these are about 2000 years old. Many vegetable fibres have been used in ropemaking, including date palm, flax, jute, cotton and *coir* (coconut fibre). Towards the end of the 19th century, manila was introduced and this soon became the most popular fibre and led to improved processing machinery. Another hard fibre, sisal, was introduced in the early 1900s and, although not so popular as manila, was a

Right: a large braiding machine making a nylon rope with an inner and outer sheath structure. The rope is designed so that loads are equally shared between sheath yarns and core yarns.

Below: making rope from coconut fibre in south India.

very useful alternative during and after the two World Wars. Sisal has been used extensively to make baling twine for hay and straw, although increasing quantities of synthetic twine are now being sold for this job. In the 1950s, nylon and the polyester Terylene [Dacron] were found to be excellent synthetic fibres for ropemaking, and these are the strongest materials available for normal commercial purposes. Another popular synthetic was polyethylene, especially during the 1960s, but although this is still used in large quantities, especially in the fishing industry, polypropylene has since been developed and is now the more popular material, generally being stronger and cheaper than polyethylene. Experiments have recently been carried out using glass fibres and carbon fibres, and a new aromatic polyamide fibre is currently being investigated, but ropes made of these materials are not available for general use.

MODERN ROPEMAKING PROCESSES Natural fibre ropes and certain synthetic ropes are made from yarn which has been spun from raw fibres. The material is fed into a goods or hackling machine which combs the fibres with steel pins and produces a coarse *sliver*. The process is repeated on other similar machines, the spacing between the pins being reduced each time so that the sliver is brought into a more regular form. The sliver is then condensed in drawing machines before being spun into yarn and wound on to bobbins. The resulting yarn can have a right handed ('Z') or left handed ('S') twist.

Synthetic yarns, such as nylon and polyester, are extruded by the manufacturers in the form of continuous filaments, and these have to be built up to a suitably sized composite yarn by plying together a number of the filaments. This is known as doubling or throwing. The yarns are then set on to a creel (a frame holding bobbins of yarn) for forming the strand of a rope. For a conventional three stranded, right hand rope, Z yarns would be used. The ends are fed through a register plate which ensures the correct formation of the strand, keeping the inside yarns located in the centre, thus obtaining optimum strength and reducing strand failure. The yarns then enter a compression tube which locates the point of twisting and helps to form a smooth strand having a compact cross-section. The formed strand is drawn through the tube and S twist is applied, the opposite direction to the yarn twist. On 'house' machines the strand is hauled by capstans and wound on to strand bobbins. Once these bobbins are filled with strand they are transferred to the closing machine. They are mounted individually at one end of the closer and the strands are brought together at the other end, a laying top being used to fix the point of twist. The rope is then passed through a die. Now Z twist is applied and the rope is closed, the finished rope being drawn through again by means of capstans or haul pulleys and wound on to a reel. To assist the closing process, the individual strands receive extra twisting to compensate for the loss of turn by the laying into rope. When the strand bobbins are empty, the finished coil is 'struck' and supported

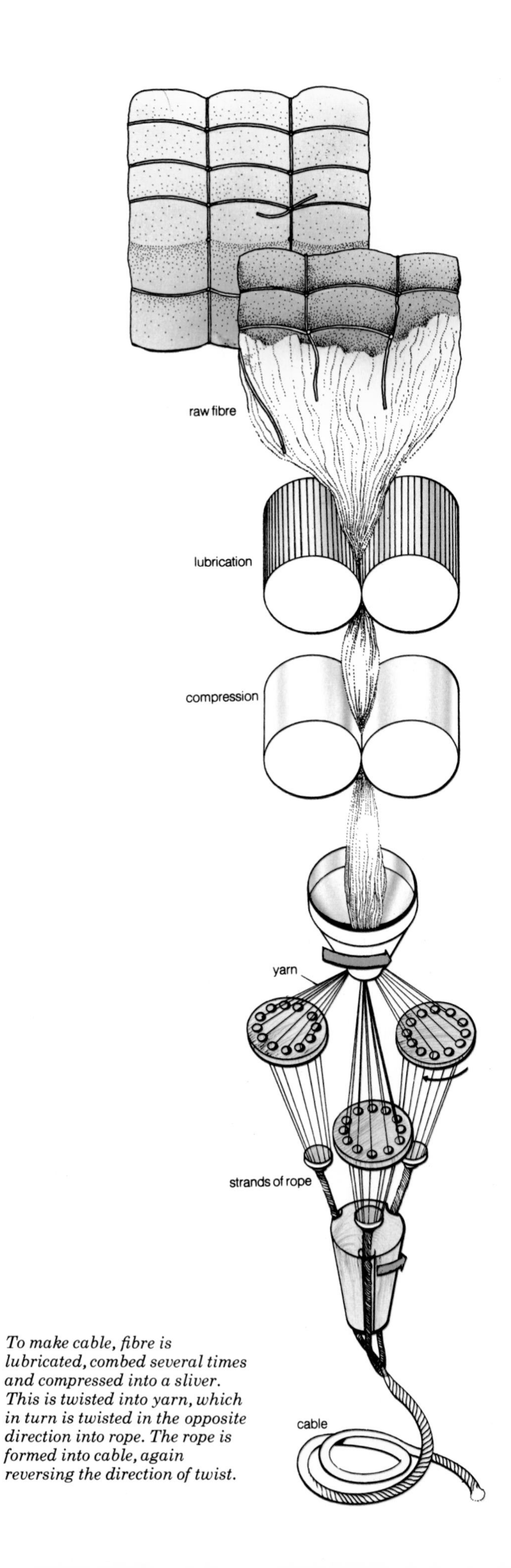

To make cable, fibre is lubricated, combed several times and compressed into a sliver. This is twisted into yarn, which in turn is twisted in the opposite direction into rope. The rope is formed into cable, again reversing the direction of twist.

by tying bands, usually of strand or smaller rope.

Another method of making rope is by means of a ropewalk, which is a path along which rope-forming strands are laid. The strands are drawn out by a traveller and, when they have been drawn to the correct length the rope is closed by means of a top-cart, which is returned from the traveller end to the register plate end at a regulated speed.

STRUCTURE OF ROPE The amount of twist in a rope is called the lay, a hard laid rope having more twist put into the strand and rope than a soft laid rope. A three stranded rope is known as plain or hawser laid, and a four stranded rope, useful for rope ladders, is known as shroud laid. Cable laid ropes are constructed from three or more stranded ropes, and the twist direction is again reversed; in other words, Z twist ropes would be cable laid with S twist.

In recent years, large plaited and braided ropes have been developed for mooring ships, and these reduce wastage by eliminating kinking. The most widely used of these ropes, called cross plait, is one in which eight strands are formed: four S direction using Z yarn and four Z direction using S yarn. These are produced on normal stranding machines, and a special closing machine then plaits the rope in pairs of strands—two

Above: sisal fibres, much used for cordage, being processed after scraping from the plant leaves.

Below: register plate of a large ropemaking machine.

Z strands followed by two S strands and so on. This construction produces a very flexible rope and is particularly suitable for nylon and polyester ropes where wastage is expensive.

Perhaps the most significant feature of modern ropemaking is the choice that is now available. A specialist rope can usually be made to suit a specialist purpose, whether this be a 'single point mooring' for a 200,000 ton oil tanker, nylon ropes for use in the North Sea oilfields, spun polyester ropes for high performance in sailing and yachting, or continuous filament polyester rope for towing in confined spaces.

SAUSAGE making

The first commercial sausage making in Britain was to be found in the pork butcher's shop. Cutting pig meat from the carcase, he tried to satisfy customer preference for lean meat, which resulted in residual trimmings which were too fatty to be used for mince. Combination of these trimmings with a starchy material, such as bread, and with salt and flavourings, gave a mix which on cooking had acceptable flavour and texture. After enclosing the mix in a *natural skin*, the sausage rope was twisted to give the traditional 'string' of sausages.

The British fresh sausage is subject to control by

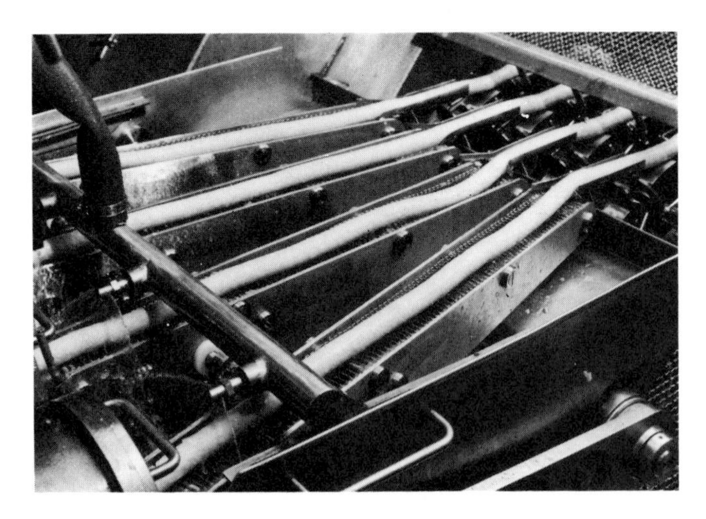

Above: four continuous 'ropes' (lines) of sausage going through an automated sausage making machine.

Above right: the bowl chopper which prepares the meat. The bowl rotates, carrying meat into the blades.

Below right: traditional sausage making in France.

Below: process diagram. Meat and other ingredients are prepared separately and blended in the bowl chopper. The collagen casing simplifies automated production.

legislation. A pork sausage must contain at least 65% meat; products with a combination of meats (pork, beef and occasionally mutton) must be at least 50% meat. 'Meat' comprises both lean and fat and at least half the total meat in a sausage must be lean.

The basic principles of sausage making apply whether it is the butcher making up a small batch or the sausage industry making up 400 lb (181 kg) at a time.

PROCESSING Lean meat, originally live muscle, is composed of fibres (the grain of the meat) lying side by side, whose length and toughness depend on the function of the muscle during the life of the animal. In order to break down large meat pieces to a particle size suitable for sausage making, a cutting operation is used rather than mincing, because knife cutting with a keen blade ruptures less of the tissue than mincing, which is shearing action under pressure, crushing the meat and tearing it apart. Cutting the meat in the presence of salt tends to seal the cut surfaces as they are exposed; ice and water added at this stage will maintain a low temperature and ensure that the salt is evenly distributed to all the meat particles.

The machine used is the bowl chopper. Vertical curved blades rotate about an axis close to the inner surface of a rotating bowl. The rotation of the bowl carries the meat into the blades.

Chopping the meat with salt and water also tends to dissolve some of the meat protein; this solution will gel when it is heated. Thus the particles of meat are coated with a salt-soluble protein solution, which coheres when the sausages are cooked, reducing loss of juices and fats.

To spread out the richness of the chopped meats and to retain juices liberated on cooking, a starchy binder is added, which may be bread, rice or, more usually, the crumpled, baked, unleavened 'biscuit' called rusk in the food trade. Flavourings are added to the chopping at any stage when uniform distribution can be obtained. Food regulations permit the addition of a preservative, sulphur dioxide, at a level of not

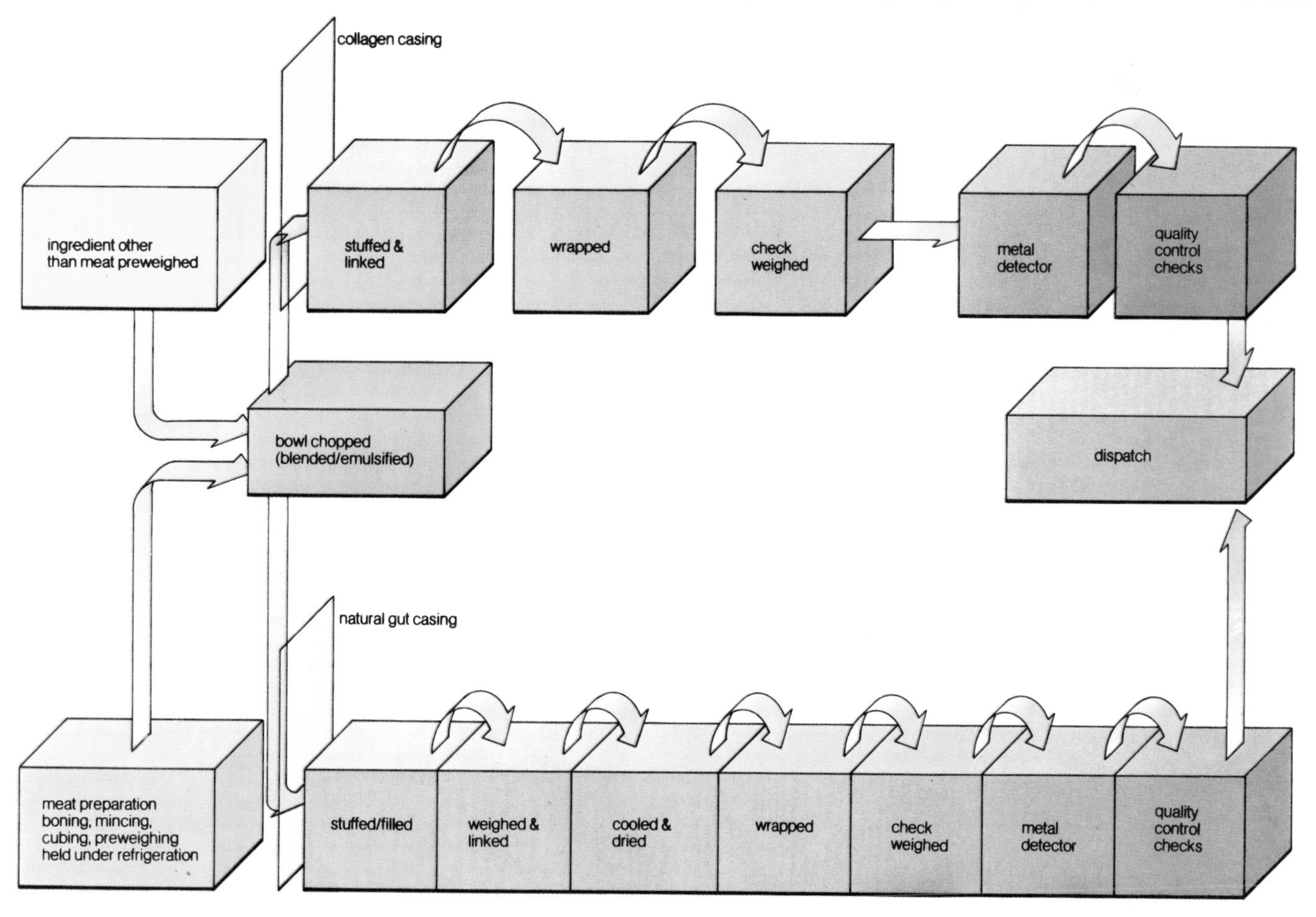

more than 450 parts per million to retard spoilage. Fresh sausages should be kept at about 7°C (44°F) or lower.

CASINGS The sausage mix is extruded into a skin. Whether the skin is natural or artificial (manufactured), it is of animal origin. The natural skin is the cleaned and prepared supporting membrane of animal small intestine, a tube of fibrous connective tissue. For thick sausage it is prepared from the pig and for thin sausage from the sheep. The robustness of natural casing comes from the fibres of the protein connective tissue, called collagen; in one form or another, collagen is also the raw material of edible 'artificial' casings.

PACKAGING The rope of sausage mix in its casing is portioned into short lengths to the weight for packing, with extra for drying and cooling loss. Hand linking by twisting fashions the length to an open bundle. This bundle is conveyed through a cool drying atmosphere to give the sausages a desirable sheen or bloom. Bundles are weighed and adjusted to pack weight plus a safety margin, and wrapped by hand using printed transparent film.

An efficient in-line system of continuous mix production fed by metered ingredients is being developed. Mechanized filling out and forming systems are already in use for sausages with artificial casings; this system is not strictly speaking continuous because the units produced are of finite length.

A more revolutionary continuous sausage-forming installation co-extrudes the mix with a coating of casing precursor, which is converted on the rope to the artificial casing.

SCREWS

As a means of fastening parts together and allowing them to be separated without damage, the use of screws is one of the oldest, cheapest and simplest methods of all. In their crudest forms, screws have been used since ancient times; however, it was not until the Renaissance that metal screws in any way resembling those we know today began to appear in quantity in such manufactured items as clocks and weapons. The reason for this was the great difficulty found in cutting or forming the screw thread on the metal blank. Early screws had to be made by tedious hand filing or, sometimes, by soldering a wire in a spiral form around the rod. This meant invariably that no two screws were ever alike.

Although Leonardo da Vinci (1452-1519) produced designs for screw cutting machines, the invention of the first practical screw cutting machine was credited to a Frenchman, Jacques Besson, in 1586. The next important stage in thread cutting development occurred in 1760 when two English brothers, Job and William Wyatt, took out a patent for a method of cutting screws at the previously undreamed of rate of nearly ten a minute. Subsequently, the Wyatt brothers set up a factory for screw manufacture and this was

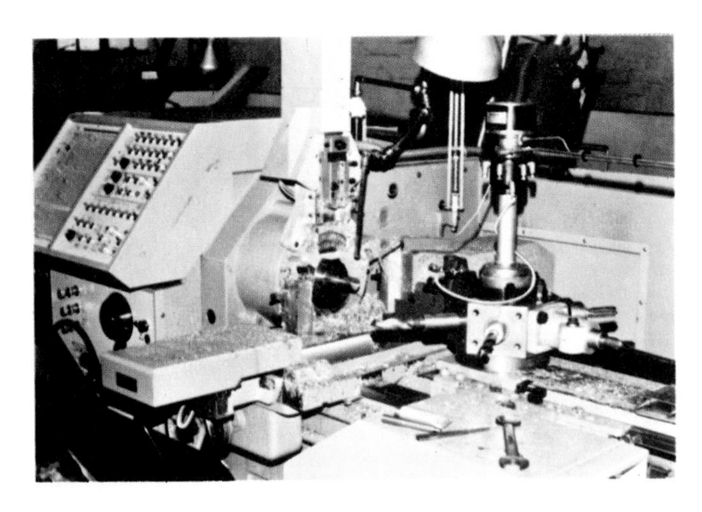

one of the earliest known examples of mass production. Henry Maudslay, another Englishman, between the years 1800 and 1810 devised the exact and scientific technique still used by today's engineers, building the first screw cutting-lathe.

At about the same time in the United States, David Wilkinson designed and built his screw cutting lathe and later, in 1845, Stephen Fitch built the first turret lathe, specifically to produce screws for an arms contract. (A turret lathe is a cutting machine with several specialized tools mounted in a revolving turret.) Shortly after the American Civil War, Christopher Walker invented the completely auto-

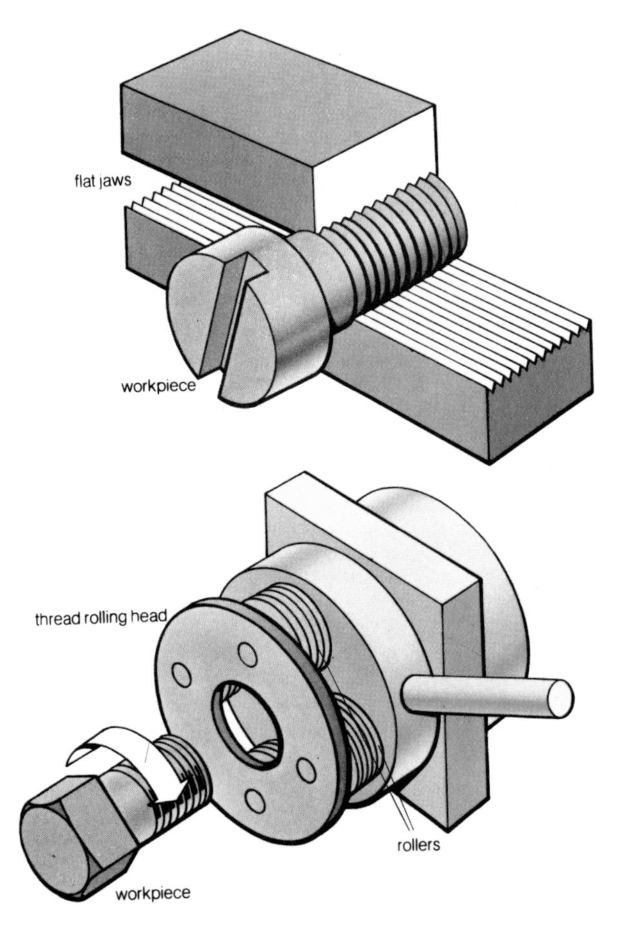

matic lathe solely to make screws to improve the assembly of his repeating rifles. As with many other inventions, the development was largely inspired by the production of weapons.

While machine cutting of screw threads is still employed today in many applications, for mass production it has been largely superseded by a process called thread rolling for the production of screws and other threaded fasteners. Thread rolling is attributed to an American inventor, William Keane, in 1836, but it had little success at first because the grade of iron used to make the screw blanks tended to split under the pressure of the rolling dies.

The production of screws by machining is a fairly limited technique, employed only on fasteners of unusual design or those too small or too large to be processed by other methods. Both standard automatic lathes and special automatic screw machines are employed, using hexagonal or round stock in cut lengths. Machining of screws is wasteful of metal, relatively slow and, while finish is invariably excellent, production costs are high.

Screws are produced in a wide range of sizes, head styles and materials, all of which are determined by their applications. Until quite recently, the thread forms were equally numerous and this created problems both in standardization and manufacture. Accordingly, in 1966 the International Standards Organization proposed the restriction of threads to ISO metric and inch, coarse and fine pitches in what is known as the preferred range of sizes. These proposals have been generally adopted throughout the world and the advantages to manufacturing are considerable. The pitch, also called basic pitch, is the distance between a point on one thread to the corresponding point on the next thread. Pitch is also taken to mean the number of threads per inch or per centimetre. For example, the Unified Standard pitches for a $\frac{1}{4}$ inch screw are $\frac{1}{4}$ 20 (coarse), $\frac{1}{4}$ 28 (fine) and $\frac{1}{4}$ 32 (extra fine).

Screws are made from specific heading qualities (suitability for heading—see below) of low to medium carbon steel wire, stainless steel, nickel alloys, brass and aluminium alloy. While some of these materials are more difficult to process than others, the production methods remain much the same. But whatever the production method used, material quality is important so as to reduce the incidence of cracks and other defects in the finished product.

COLD HEADING Cold heading is a widely used method of forming or *upsetting* a head of predetermined size and shape on one end of a cut blank of rod or wire. Production is a relatively high speed continuous process; the wire is fed from a coil mounted on a *swift*, or *pay-off* unit. The end of the coil may be welded to the end of the next one. The wire passes through a pre-straightening unit into the machine where, in a predetermined sequence, it is cut to the correct length and, depending upon the number of blows required to form the head, it is punched into a tungsten-carbide die so that the head takes up the required shape, for example, hexagonal, round, recessed, and so on. After forming, the blank is automatically ejected into a receiver for further processing.

The two basic types of cold heading machine are those fitted with split (open) dies and solid (closed) dies. Split dies are used for making screws or parts with wide tolerances and greater than average lengths; solid dies are designed to achieve greater

Top left: program-controlled thread cutter for small batches. Note the revolving capstan.

Bottom left: two methods of thread rolling: the reciprocal method, where the blank is revolved between flat dies; and the cylindrical method, where it is turned within a circular arrangement of rollers.

Below: the next five pictures show the mass production of screws. First, a cold heading machine; it is fed with wire, makes two blows on the end to form the head, then cuts the required length off. A different method is shown in the second picture; a multi-station transfer heading machine, which transfers the workpiece from station to station for various stages of formation.

Above: head slotting machine; the blanks are set around the edge of the wheel.

Right above and below: feed to thread rolling machine, and the machine itself.

accuracy (closer tolerances) and also to allow a degree of extrusion, that is, a reduction in the shank diameter of the screw blank. This extrusion, which is achieved within the die, is necessary so that when the threads are rolled into the metal at a later stage of production, the correct major diameter is accomplished. For screw manufacture the cold heading machines are also classified as either single blow or double blow. The single blow header applies the heading punch once for each revolution of the machine's flywheel to produce one screw for each stroke of the punch ram. The double blow header is fitted with an indexing head containing two punches, each of which is applied once during the cycle. Thus, two strokes of the ram are required to produce one screw blank. As an alternative, the die itself may be indexed automatically instead of the punch. Production speeds, depending on the diameter and length of the screw, can range from about 100 to 550 parts per minute.

A more advanced type of cold header is the transfer machine, which is rather like a series of single blow, single die headers, each linked to the other by means of a transfer mechanism. In this machine the blank is ejected from the die after each blow, to be transferred to the next station for progressive forming. One of the advantages of this type is that trimming, reduction and pointing are carried out in the one machine. By its very nature cold heading is a noisy operation and machine designers and manufacturers are incorporating soundproof hoods in their latest designs.

There are many advantages of cold heading, including the high volume of production achieved and the complete elimination of material wastage. (Because of this, cold heading is called a 'chipless' method of production.) Also, cold heading causes the metal grain flow to follow the contours of the head, thus avoiding stress, particularly where the underside of the head joins the shank. The process allows the use of low carbon steels for highly stressed application because the cold working which takes place during heading actually improves the mechanical qualities of the metal.

Disadvantages of the method are generally restricted to size. For example, very small or very large screws tend to be uneconomic, the former due to handling problems and the latter because large screws demand large and powerful machines.

The intermediate stages between cold heading of the screws and the thread rolling operation may include slotting of the head, trimming and pointing.

THREAD ROLLING Thread rolling is also a cold forming process. The thread form is impressed in the screw shank by rolling it in a single operation under controlled pressure between two hardened dies having the reverse profile of the specified thread. The indentation of the die thread crests causes the metal to fill the area between the thread flanks by plastic deformation. Since no metal is removed from the screw blank, but is only displaced, the blank diameter on which the thread is rolled must be slightly undersize, about equal to the thread pitch diameter.

The pitch diameter is also called the simple effective diameter. It is the diameter of an imaginary cylinder, the surface of which would pass through the thread profiles at such a point as to make the width of the remaining groove equal to one-half the basic pitch. On a *perfect* thread (where the groove and the thread are the same size) this will be the point at which the widths of the thread and the groove are equal.

There are three types of thread rolling processes, the flat (reciprocating) die, the centreless cylindrical die and the planetary rotary die. The first two processes have production speeds in the range 60 to 250 parts per minute, depending on blank diameter, while the planetary die type achieves speeds of between 60 and 2000 parts per minute. In the reciprocating die method, there are two flat dies, one of which is stationary while the other reciprocates, rolling the screw blank between them. With centreless cylindrical dies, there are two or three round dies and the blank is rolled between them. In the planetary die method the blank is held while the dies roll around it.

In most thread rolling operations, the screw blanks are fed automatically to the dies down a guide chute from a vibrating hopper feeder which ensures that the blanks are correctly presented and at the correct feed rate.

Although the point is frequently argued, the rolled screw thread is superior to the cut thread, since it has the same characteristics as a cold headed product in that the fibres of the metal follow the contour of the thread and are not discontinued or severed as in the case of cut threads. The roots of rolled threads are stressed in compression, thus improving the fatigue strength, particularly in medium carbon steels. Moreover, rolling between dies leaves a thread with smooth burnished root and flanks free from cutter marks that may create stress.

SPORTING GOODS

The origins of some games are lost in the past (golf, for instance, appears to have originated in the 15th century), but others were developed over a short period in more recent times. The adoption of new materials and manufacturing techniques for sporting goods is severely limited by the rules governing the games which are often designed to preserve traditional aspects. For instance, the resilience of golf balls made in the USA (and hence flight distance) is strictly limited, and such a rule is likely to be introduced worldwide.

Wood and leather, being natural materials, have been used for many years for golf, tennis, cricket equipment and so on, and there is still a considerable requirement for these materials. Newer materials such as high strength alloys, rubber, plastics, glass and carbon fibre are now also used in large quantities.

TENNIS BALLS Tennis (or lawn tennis to give its correct name) was developed from real or royal tennis, a game going back to the 14th century. Lawn tennis was perfected in the 1870s and was based on a ball made of rubber which bounced considerably better than the cloth-stuffed ball used for real tennis. Covering the rubber with felt was found to improve its wearing qualities. The felt seams were originally joined by stitching, but this was replaced by rubber cement in the 1920s.

The modern manufacturing method consists of first mixing rubber with special ingredients (clay to reinforce, sulphur and 'accelerators' for vulcanization) in an 'internal' mixer which consists of a pair of intermeshing rotors rotating within a chamber, which continuously shear the rubber-powder mixture until it is homogeneously blended. The mixture or compound, which is of putty-like consistency, is then forced by hydraulic pressure through the circular die of an extruder to produce a continuous rod, which is chopped into pieces or *plugs* of identical size and weight by a

Above: roughing out billiard balls from ivory blocks in France in the mid-19th century. Today they are made of plastic.

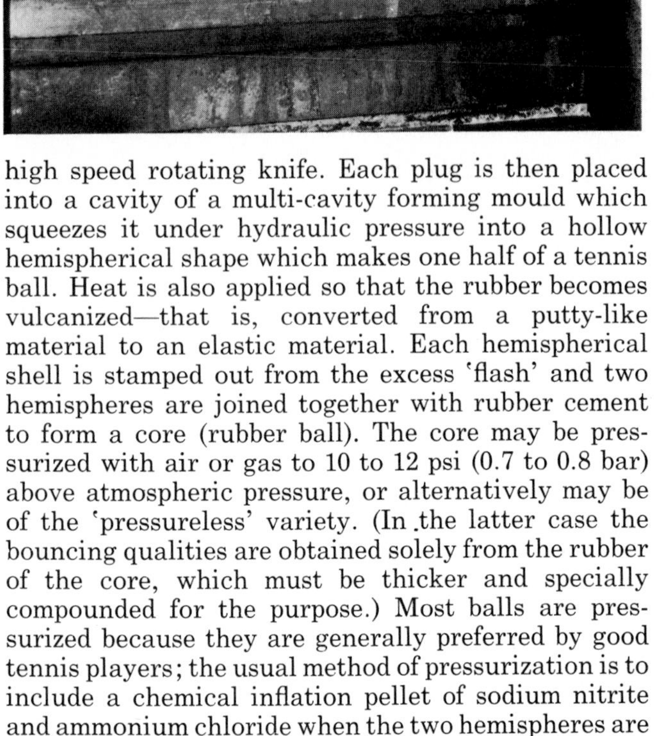

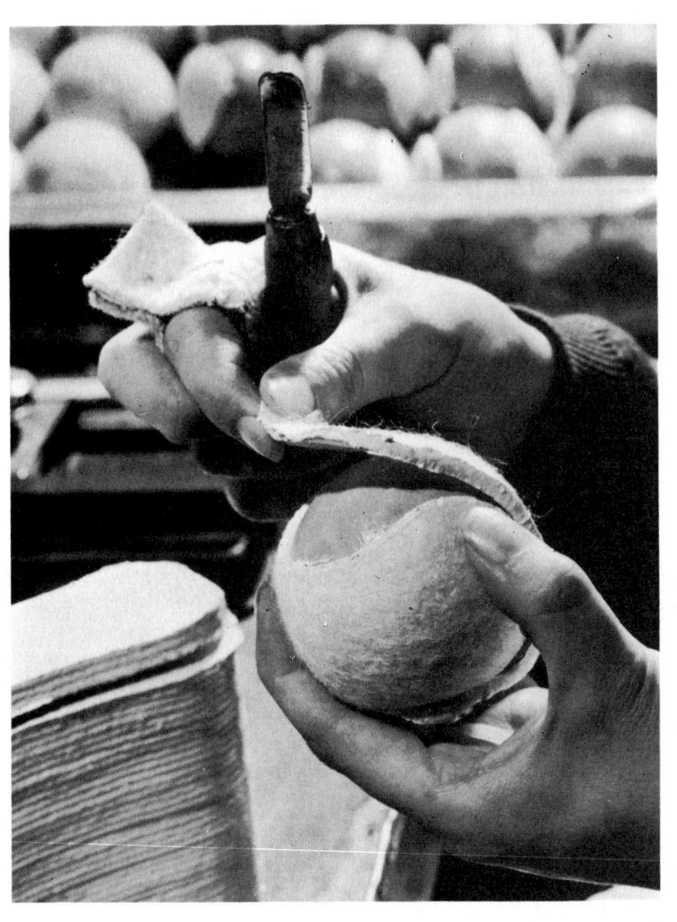

high speed rotating knife. Each plug is then placed into a cavity of a multi-cavity forming mould which squeezes it under hydraulic pressure into a hollow hemispherical shape which makes one half of a tennis ball. Heat is also applied so that the rubber becomes vulcanized—that is, converted from a putty-like material to an elastic material. Each hemispherical shell is stamped out from the excess 'flash' and two hemispheres are joined together with rubber cement to form a core (rubber ball). The core may be pressurized with air or gas to 10 to 12 psi (0.7 to 0.8 bar) above atmospheric pressure, or alternatively may be of the 'pressureless' variety. (In the latter case the bouncing qualities are obtained solely from the rubber of the core, which must be thicker and specially compounded for the purpose.) Most balls are pressurized because they are generally preferred by good tennis players; the usual method of pressurization is to include a chemical inflation pellet of sodium nitrite and ammonium chloride when the two hemispheres are glued together. When heat is applied, the pellet decomposes, releasing nitrogen gas.

The felt or melton with which a tennis ball is covered is a high quality cloth of wool and nylon. The weft (transverse) yarns are made from a wool and nylon mixture and these are woven into the warp (longitudinal) yarns, which are cotton, in such a way that the weft appears predominantly on one side of the cloth. This surface is subject to a teaseling or 'raising' operation to produce a hairy surface which is then consolidated by fulling—a process in which the natural felting properties of the wool are exploited by working the cloth in a soap solution to produce the necessary surface texture.

The melton is coated with rubber solution on its reverse side and is then cut into dumb-bell shapes, two of which are used together to completely cover the surface of the core. The dumb-bell 'covers' are applied by hand and the degree of stretching is carefully controlled so that an exact fit is obtained. Rubber cement applied to the edges of the dumb-bells becomes vulcanized in a further moulding operation in which the ball is heated in spherical moulds. A steaming operation raises the nap and the ball is tested for deformation under load, so that balls are matched together before they are packaged.

Tennis balls must fulfil a rigid specification determining size, weight, rebound and compression, and careful quality control must be carried out at all stages of the manufacturing operation.

GOLF BALLS Golf balls were originally made by stuffing feathers into a hand stitched leather case under considerable pressure. From 1850, they were made from solid gutta percha (a substance related to rubber, coming from the juices of Malayan trees) until the rubber-cored ball was developed about 1900, and this rapidly became accepted after it was used by the winner of the Open Championship in 1902.

The rules of golf lay down only the maximum weight (1.62 oz) and minimum size for a golf ball: 1.62 inch

(4.11 cm) in England and elsewhere except the North American Continent, where it is 1.68 inch (4.27 cm) and a maximum resilience is also specified.

A golf ball consists of 3 main components which are a centre (usually liquid or resilient rubber) around which are placed windings of highly elastic rubber thread and a cover to protect the thread and to incorporate the 'dimple' pattern.

A considerable amount of the mass of the ball is concentrated in its centre, which must allow the windings to distort readily when the ball is struck by the club, so that the subsequent rapid recovery of the highly tensioned thread creates high ball velocity. Because the centre must, however, absorb a minimum amount of energy at ball-club contact, the liquid centre has been used in premium balls for many years.

One way of making a liquid centre is as follows. Fine clay is mixed into water and glycerine and the mixture is measured out into hemispherical cavities in rubber moulds. Because of the thixotropic nature of the mixture (a tendency to become thinner when stirred or shaken but having a high viscosity when undisturbed—a property of certain fluids and plastic solids), the rubber moulds can be brought together in a vertical orientation so that spheres of paste are produced. The moulds are refrigerated to freeze the spheres of paste and these are then removed and coated with rubber and subjected to a hot moulding operation to vulcanize the rubber coating. The result is a small, heavy, deformable sphere of liquid inside a rubber envelope.

Rubber thread is produced by mixing rubber with special ingredients to obtain a highly elastic sheet after vulcanization—this process being carried out by winding the sheet on to large drums which are put into steam-heated chambers. The sheet is then passed between multi-knife cutters to produce rubber thread of dimensions approximately .060 × .020 inch (1.52 × 0.51 mm).

The golf ball core is made by stretching the rubber thread to about 900% and winding it on to the centre by means of a core winding machine. The core is held between high speed rotating rollers which both rotate the core and also allow it to turn about any axis. The operation of the machine causes the stretched thread

Left: stages in the manufacture of tennis balls. First, circular pieces of a putty-like compound are vulcanized into half-round shapes, which will be cemented together around an inflation pellet. Next, the cover is hand fitted and cemented on. Finally, a moulding process ensures a truly round shape.

Below: this machine turns the centre of a golf ball – a sphere of liquid enclosed in a rubber envelope – while rubber thread, stretched to 9 times its original length, is wrapped around it.

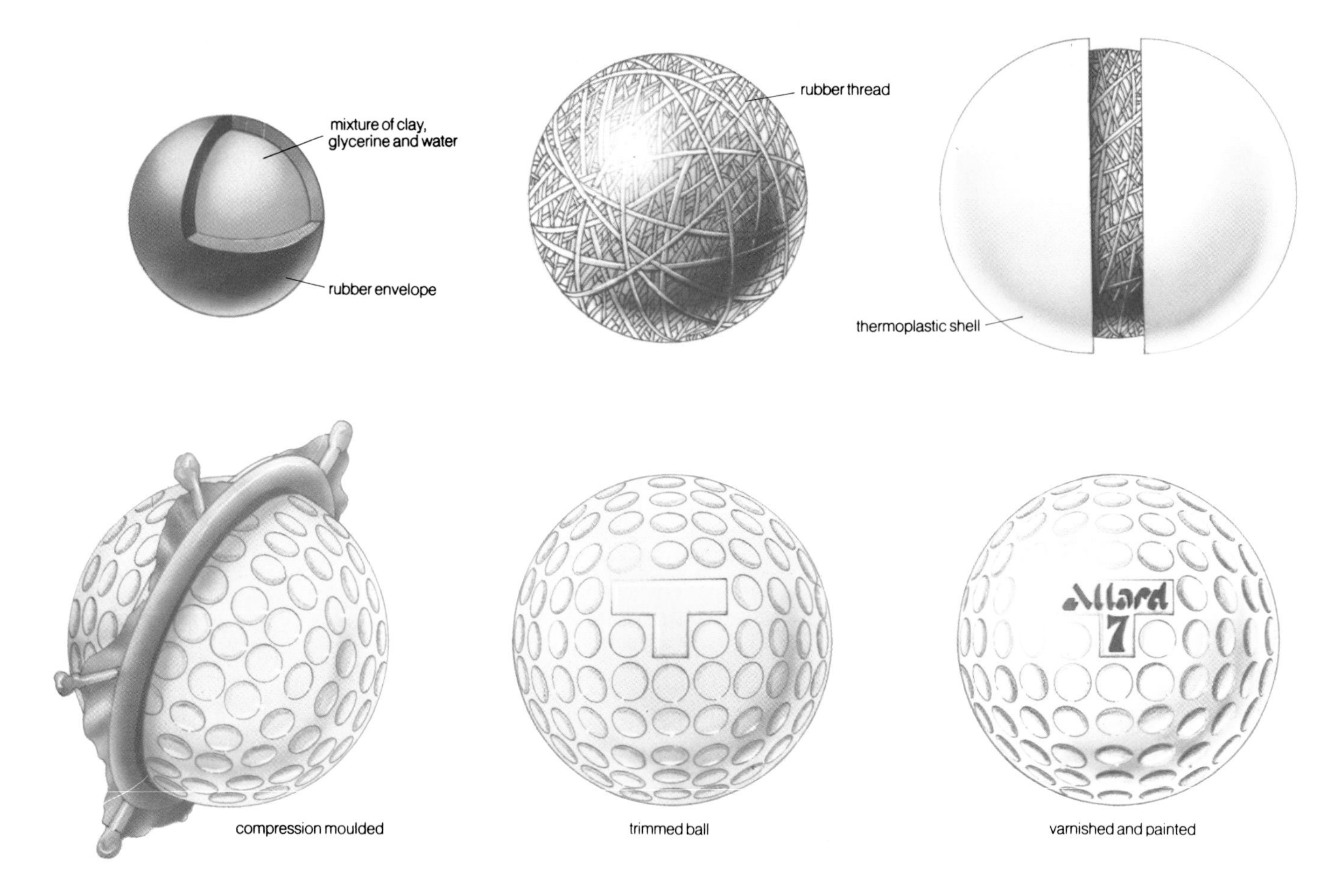

Above: the small, heavy deformable centre of a golf ball is made as shown here to absorb a minimum of energy when struck. The compression-moulded cover has dimples which are aerodynamically important for flight distance.

Right: a wood tennis racket is made of strips bent cold and glued with urea formaldehyde adhesive in a forming jig under hydraulic pressure, then heated to set the glue. These 'heads' are wide enough for three rackets to be cut from each.

to be wound on to the smallest diameter of the core at any one time, and the core continuously reorientates itself and so becomes spherical. The machine cuts out automatically when the core grows to the correct size.

The material to form the cover of the ball is a thermoplastic material which is formed into hemispherical shells on an injection moulding machine in which the hot, plastic material is forced under pressure into cold moulds. Two shells are placed around each core and the assemblies are inserted into precision dies in a compression moulding press, which moulds the cover material on to the cores under the action of heat and pressure. The dimple pattern is moulded into the surface of the ball by the profiled surface of the die.

The moulds are cooled and the balls extracted and accurately trimmed free of excess material. Pre-paint treatment follows and the balls are spray-painted on machines adapted for painting spherical objects. The balls are matched for compression (that is, deformation under load), the low compression (or high deformation) balls being segregated as lower grade. The balls are identified by a stamping process and finally coated with clear polyurethane lacquer.

The dimple pattern moulded into the ball surface has a very important function and the size and shape of the dimples are critical. The golf club is designed to produce a 'backspin' to the ball at contact—so that although the ball is projected forward, it is also caused to spin about a horizontal axis so that the top of the ball is moving against the direction in which the ball is travelling. The air flow over the top of the ball is therefore speeded up and that below the ball is retarded. This produces a local reduction of air pressure immediately above the ball and an increase in that immediately below the ball such that a resultant upward force or lift occurs. The dimple pattern controls the degree of lift generated by influencing the interaction between the ball surface and the air flow, and it also affects the 'drag' experienced by the ball in moving through the air. The distance the ball travels through the air is therefore directly dependent on the dimple pattern and a considerable amount of experimentation has been carried out over the years to produce the most efficient pattern in terms of flight distance.

TENNIS RACKETS A tennis racket is not subject to any specification by the rules of lawn tennis but appropriate shapes, sizes, weights and balances have become established by common practice.

Until the 1920s, racket frames were made by bending single sawn sticks of ash to the familiar racket shape after they had been softened by steaming. Subsequently, similar types of frame were produced by steam bending several thinner sticks and gluing them together to produce a plied frame. In the 1930s and 40s a process was developed for producing frames from even thinner sticks or veneers which could be bent cold, and strength was improved by adopting urea formaldehyde glues. This is the basis of the laminated wooden frame which we know today.

Various woods are used in each frame. The basic strength of the frame is provided by ash and beech, but the throat or wedge area is usually either sycamore or mahogany. Lightweight obeche is used as a spacer in the handle region, and hickory may be used to give strength and wear resistance on the outside of the frame. In some cases, a wood such as walnut is used for its decorative appearance.

The sticks or veneers are obtained from specially selected logs by sawing, slicing or peeling. Peeling is more efficient in yield of wood. In this operation, the log is steamed for a considerable time, and is then rotated about its axis while a continuous veneer of wood is pared away by a knife blade which contacts the log along its whole length. The veneer is cut into strips parallel to the grain, which are bent to form the basic 'keyhole' contour of the frame, so that the grain of the wood is used most effectively for strength and rigidity. Shaped pieces for the throat and handle are specially profiled and all the components are coated on mating surfaces with urea formaldehyde glue. They are then assembled into a bending jig and the characteristic racket shape is produced by the application of hydraulic pressure, which bends and consolidates the component parts around a former. Clamps are attached so that the formed bend may be removed from the jig and passed through ovens to cure the adhesive.

In modern manufacturing practice, up to three racket frames can be cut longitudinally from one bend. Handle pieces are then glued on together with reinforcement in the shoulder area and the frame is then subject to a succession of shaping and sanding operations. Stringing holes are produced on a multispindle drilling machine, and subsequent countersinking and grooving operations are carried out largely by hand. The application of transfers, painting and lacquering complete the process of manufacture.

Wooden squash and badminton rackets are manufactured in a similar manner, but often such rackets incorporate a steel shaft for improved lightness and strength.

While tennis rackets are also made from steel and aluminium alloys, composite metal and glass fibre and carbon fibre constructions, such rackets are considerably more expensive than wooden ones. Wooden rackets are preferred by the majority of the top-class tournament players and outnumber the other types by

Left: multi-spindle drilling machine used to make all the stringing holes at once.

Below: stringing is done with the help of a machine. Strings for tournament tennis rackets are still made of gut, but most others of nylon, Tension is up to 27kg (60lb.).

Right: top-quality sporting goods still require a lot of old-fashioned hand craftsmanship. Here a cricket ball is being hand-stitched.

about 10 to 1. Metal-headed badminton rackets have substantially replaced wooden ones due to their improved strength and lower weight but in the case of squash rackets, only wooden headed frames are allowed by the rules of the game.

Racket strings are made from natural gut (produced from the intestines of sheep) and although such strings are preferred for top-class tennis, strings made from nylon filaments by a spinning or braiding process are popular because of lower cost and improved durability. Typical stringing tensions are 55 to 60 pounds (25 to 27 kg) for a tennis racket, 35 to 40 pounds (16 to 18 kg) for a squash racket and 25 to 30 pounds (11 to 14 kg) for a badminton racket.

GOLF CLUBS Golf clubs are subject to certain specifications regarding their general constitution designed to 'preserve the character of the game' and certain aspects such as face markings and grip shapes are closely specified.

The rules allow the golfer to use a set of not more than fourteen clubs; these would typically consist of 4 woods, 9 irons and a putter. Clubs vary in weight (mainly determined by the head), length (determined by the shaft) and loft, which is the angle the face of the club makes with the vertical in the address position. Clubs get shorter, heavier and more lofted in their numbered sequence.

The heads of wooden clubs were once largely made from persimmon but nowadays are made from specially shaped laminated blocks of maple, shaped so that

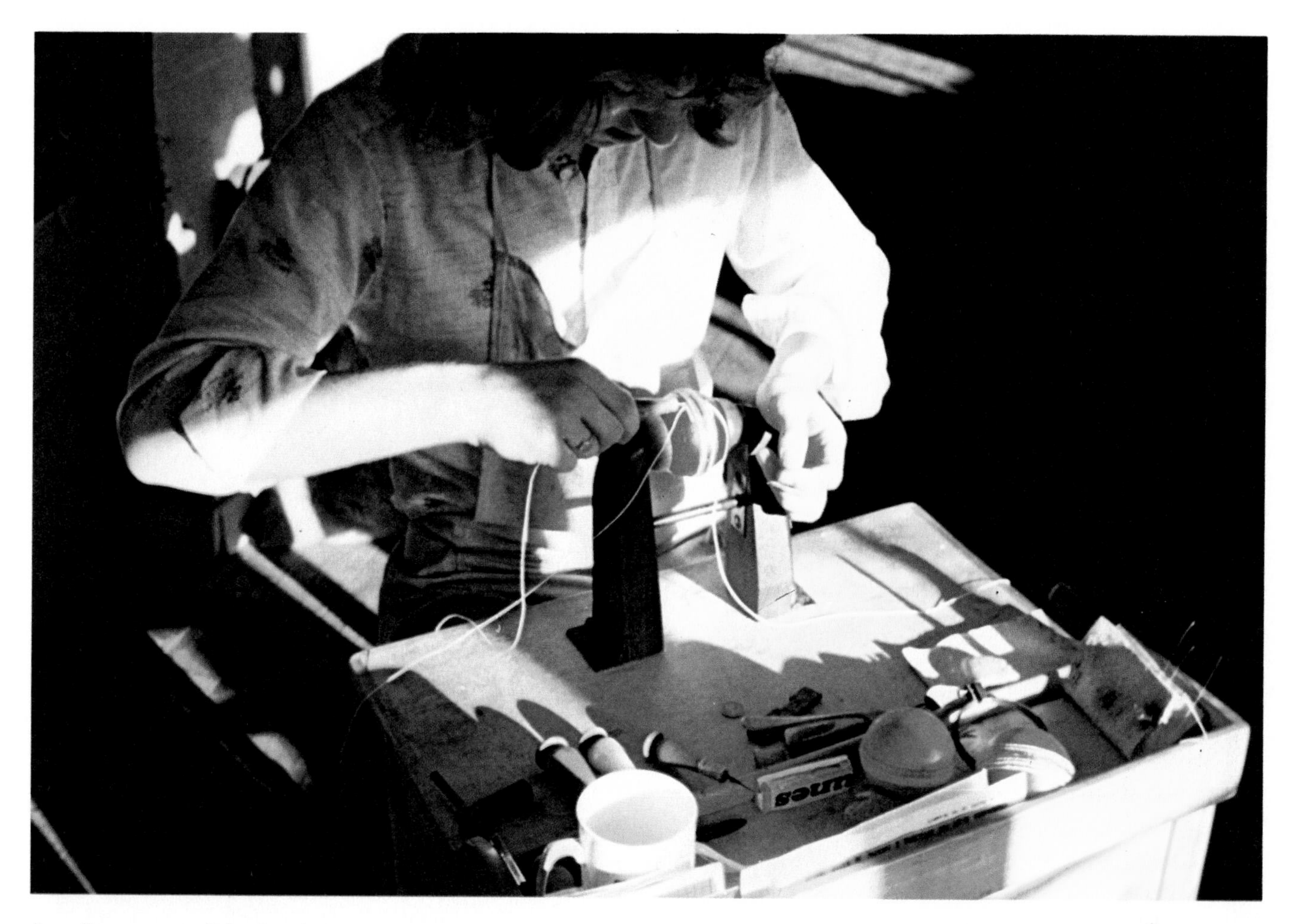

the alignment of the laminations changes to maximize the directional strength in the *hosel* area, where the shaft is inserted. The heads are turned on a copy lathe to produce the rough shape characteristic of a golf wood, and areas are routed out on the hitting surface (face) and base (sole) to allow the fitting of a plastic facepiece and metal soleplate respectively. The weight of the wooden head is adjusted by adding lead to a hole drilled in the sole before the baseplate is screwed into position. The wood surface is sanded and painted after carefully shaping the club face.

Iron heads were originally hand forged but now are either drop-forged or cast by an investment casting process. In this process the metal is poured into a ceramic mould which is often made from a wax master (lost wax process). Mild steel is primarily used for forged heads, but stainless steels are common in castings. Castings have the advantage that they can be produced to closer tolerances than forgings and allow more design freedom for weight distribution and intricate detail. Additionally, the hole for shaft attachment may be cast rather than drilled. The heads are subject to grinding and polishing operations to produce correct weight and a high surface finish, and are sometimes chromium plated.

In early days shafts were made of split ash and later hickory. Tubular metal shafts date from the 1890s but were not exploited until the early 1920s; although they gained acceptance in USA, they were not authorized for golf in England until 1929. Shafts are now produced from steel alloy (manganese-boron) in tapered form by a step-tapering process in which metal tube is forced through dies of progressively smaller diameter over reducing length of the shaft to produce the characteristic 'stepped-taper' effect. By this means the appropriate distribution of flex is obtained and the smaller diameter of the shaft where it is attached to the the head has a thickened wall for improved strength. Different overall stiffness can also be produced by changing the wall thickness of the original tube.

Golf shafts have also been made from other materials, notably aluminium and carbon fibre. The object is to produce shafts of lighter weight than steel shafts and hence produce a more efficient club, but aluminium allowed only a marginal improvement in this respect. Carbon fibre allows a 30% reduction in shaft weight and some of this weight can be transferred to the clubhead with a resultant improvement in ball velocity after impact. Because the fibres must be aligned for satisfactory longitudinal stiffness, however, torsional properties of carbon fibre shafts tend to be poor and this can adversely affect the golf shot.

The shafts are cut to the appropriate length and glued or riveted to the head. Leather or rubber grips are then fitted and the balance or swing-weight

Above: a craftsman shaping a cricket bat with a spokeshave. The blade is made of willow, which is allowed to dry naturally and then compressed for strength. The handle is made of cane laminated with rubber strips.

of the club, about a point 6 inches (15.2 cm) from the top of the grip, is measured so that sets of matched clubs can be made up.

OTHER EQUIPMENT Although the manufacture of golf and tennis balls, golf clubs and tennis rackets involves modern scientific technology, equipment for other games often requires the skill of craftsmen that has remained unchanged for over a hundred years. The manufacture of leather footballs is one example in which the only concession to modern technology is the use of sewing machines for stitching the casing, although hand stitching is still necessary to complete the work. In the case of cricket balls and baseballs the best quality ones are still hand-stitched.

Cricket bat making is another area in which the skill of the craftsman is still much in evidence. A cleft from which the bat blade is made is roughly split from willow, sawn into lengths and allowed to dry naturally by stacking in open sheds. The face and edges of the blade are then compressed to impart extra toughness. Sarawak cane is planed down and glued together with rubber insertions to form the bat handle which is finished by turning to shape on a lathe. A V-cut is made in the top of the blade and the handle end is shaped to fit accurately and is glued into position. The bat blade is then shaped by means of a drawknife and spokeshave (a two-handled planing tool for curved surfaces) to its traditional form, and finally twine and a rubber sleeve is put on the handle. The laws of the game decree that the bat must not be longer than 38 inches (0.965 m) or wider than 4½ inches (11.43 cm) at its widest part. (The American manufacture of baseball bats is simpler, but similar in its meticulousness and quality control.)

The method of manufacture of a hockey stick has much in common with that of a cricket bat in that sarawak cane is used for the handle and this is jointed to the head in a similar manner to the splicing of a cricket bat handle. The head of a hockey stick is usually of mulberry wood, which is bent to the familiar shape after steaming. The stick must pass through a 2 inch (5.08 cm) diameter ring to be legal.

SUGAR refining

Sugar cane existed in New Guinea 12,000 years ago and was cultivated in Egypt and the coastal lands of the Mediterranean about 3000 BC. A soft, sweet and thin-stemmed variety, known as 'Criolla', was introduced into the West Indies in 1493 by the explorer Christopher Columbus, who was the son-in-law of a wealthy Madeiran sugar planter. During the 18th century, sugar cane plantations developed rapidly in the colonies of many European countries as a result of the cheap supply of slave labour drawn from the African continent. The modern varieties of sugar cane are hybrids of *Saccharum officinarum*, *S robustum* and *S spontaneum* that have been bred to resist Mosaic disease (which attacks the leaves) and give high yields.

The sugar beet plant was discovered comparatively recently, originating as a weed, *Beta maritima*, found on Mediterranean sea shores. The variety cultivated about one hundred years ago contained only about 5% by weight of sucrose, but modern varieties, such as *Beta vulgaris*, contain up to 20% sucrose. Sugar beet is grown in temperate areas, particularly in Europe, and sugar cane in warmer areas such as Australia, the southern US and the West Indies.

CHEMICAL STRUCTURE Granulated sugar is over 99% pure sucrose. It can be transformed to two other sugars, glucose and fructose, by the addition of one molecule of water. This process of hydrolysis is called 'inversion', and can be accomplished by heating a sucrose solution with a dilute acid or by the action of enzymes. The resulting glucose and fructose are referred to as 'invert sugars'. The sugar beet at maturity contains virtually only sucrose but the sugar cane may contain appreciable quantities of invert sugars which, fortunately, are at a minimum at maturity.

RAW SUGAR PRODUCTION The principal difference between cane and beet sugar production is that the beet is more uniform and almost perfectly clean by the time it reaches the beet slicers. Thus the production of sugar from the cane, though similar in principle, requires a more involved process which can be used to illustrate the complexity of sugar manufacture.

Much of the world's sugar cane is harvested by hand-cutting after burning the crop to remove the leaves, but mechanical systems such as pushrake cutting and grab harvesting are coming into use. The cane arriving at the raw sugar factory is washed on conveyers to remove field mud, sand and trash. The

Below left: harvesting sugar cane near Lautoka, in Fiji. Sugar cane is the country's most important crop.

Below: unloading raw cane sugar in London from a bulk carrier. The sugar is imported from the West Indies and other sugar producing areas.

cane is nearly 90% juice and it was traditionally extracted by a combination of shredders and three-roll crushers squeezing out the juice. This process gave 95% extraction of the juice, but this has been improved to greater than 97% by the introduction of diffusion processes. One system, known as the ring diffuser, prepares the cane by shredding it in hammer mills, operating at 1000 rpm, which pound the cane with steel hammers. The shredded cane is fed to an annular, rotating perforated plate of the diffuser where sugars are extracted by recycled juice percolating through the cane at 160 to 165°F (71 to 75°C). The extracted cane, or bagasses, is removed from the diffuser by screw conveyers and can be used as a fuel in the steam-raising plant of the factory, which produces the steam used to heat the sugar and, in some factories, to drive the machinery.

Many of the impurities of raw cane juice are removed by adding a lime suspension to give a pH of 8.5 and heating to about 220°F (104°C), which is maintained for 20 minutes. (The pH scale is a measurement of acidity.) Acids are neutralized and the phosphates present are flocculated (coalesced) and adsorb colouring matter and colloids (suspended particles), which are subsequently removed by settling for about 3 hours in clarifying tanks. Next, the clarified juice is concentrated to about 65% by weight of sugar in a multi-stage evaporator system. The juice temperature in the first evaporator is about 229°F (109.5°C) and the pressure in succeeding evaporators is reduced so that the juice boils at 153.3°F (67.5°C) in the fourth effect. Vapour generated in each evaporator is condensed in the heating tubes of the next effect, thus economizing on the use of steam.

The concentrated juice is then boiled to a supersaturated (highly concentrated) solution in vacuum pans. Crystallization is induced by 'seeding' with a 'magnum' of sugar and syrup to form a mixture of sugar crystals and liquor, known collectively as *massecuite*. The massecuite is discharged from the vacuum pans at 160°F (71°C) into water-cooled crystallizers where further sugar crystals are formed by reducing the temperature to about 100°F (37.8°C) over a 48 hour period. The raw sugar crystals are separated by reheating the massecuite to 122°F (50°C) to reduce the viscosity, followed by treatment in basket centrifuges operating at 1500 rpm. The residual syrup purged from the massecuite is a dark viscous liquid known as 'blackstrap molasses' and it is used in the manufacture of rum, industrial alcohol and citric acid.

REFINING The first stage of raw sugar refining is called affination, and consists of removing the molasses film coating the crystals by 'mingling' the raw sugar in a U-shaped trough with a 75% sugar syrup. The syrup is removed by centrifuging and the affined sugar is 'melted' in pure water at 190°F (88°C) to give a strength of 66% by weight. Up to 50% of the colour can be removed by treating the refinery melt with phosphoric acid and adjusting the pH to 7.3 with lime. After heating to 195°F (90°C), flocculation of the im-

purities occurs and they can be removed by filtration or clarification. Final colour removal is achieved by adsorption in beds of granular carbon or bone char (carbonized bone particles), followed by filtration to remove the last traces of suspended matter.

The water-white sugar liquor is concentrated to 78% in a double effect evaporator and crystallization is initiated in the vacuum pan by further evaporation and seeding with fondant sugar. The sugar recovered in the centrifuges contains about 2% moisture which is removed by drying for 10 minutes in a 30 ft (9.1 m) long rotary drier with an air inlet temperature of 220°F (104°C). After cooling to 110°F (43°C) the sugar is transferred to silos where the air is kept at a relative humidity of about 60% to avoid re-adsorption of moisture.

Cube sugar can be produced by mixing dry sugar with 1% of sugar syrup and filling into the moulds of a cylinder which rotates against a stationary pressure bar to compress the cubes. The cubes are discharged on to a conveyer and dried in an oven at 140°F (60°C). Soft sugars vary in colour from light to dark brown. They have an invert sugar content of up to 6% and a moisture content of 4%.

Liquid sugar can be a pure sucrose syrup of 67.5% concentration or a mixture of sucrose and invert sugar. Some liquid sugars are blended with corn syrup. Powdered sugars, such as icing sugar, are prepared by grinding granulated sugar, and their tendency to cake is often overcome by incorporating 3% corn starch.

Left: a row of filter presses used in sugar refining.

Above: taking a test sample from a vacuum pan.

Below right: many simple tools are still made by hand. A Japanese tradesman cutting saw teeth at a market.

TOOL manufacture

Man has been making tools for half a million years; it is not surprising that the general shape and application of modern hand tools was formulated as long ago as Roman times. Most ancient tools are instantly recognizable for what they are by modern workmen. Hand tools are an excellent example of ergonomic design, since the workman knows better than anyone what type of tool is easiest to use, and workmen usually made their own tools until the 19th century.

The earliest tool was probably the axe, made of flint or stone in Neolithic times (4000 to 50C0 years BC). Europe was largely forest in those days, and early farmers changed the course of human history when they used their axes to clear the land. An early example of the specialization of tools was the development of the adze, a primitive type of chisel; in comparison with the axe blade which was securely mounted to the handle with the cutting edge parallel to it, the adze was more lightly bound with the cutting edge at a right angle to the handle. The axe was designed to split wood and the adze was invented to remove strips of wood in order to shape it.

The saw and the bow drill were developed in ancient Egypt. Saws were made of copper or bronze, and the teeth were sharpened on the back edge so that the cutting took place as the workman pulled the saw toward him, to avoid buckling the soft metal. An early advance in technology was the setting of the saw teeth in alternate directions, which carried away the sawdust and left a kerf (saw-slot) slightly wider than the saw blade, reducing friction. Early carpenters, having no screws or nails, secured their constructions with wooden dowels (round pegs). The bow drill was used as long ago as 3000 BC to do the drilling. The brace and bit, which allowed constant rather than intermittent drilling, was not invented until the 15th century, in northern Europe.

Workmen had no schools or textbooks to teach them how to make tools; each generation taught the next on the job, often composing rhymes and jingles to make their techniques easier to remember. Hand tools throughout history have survived in greater numbers than any other artifacts, because workmen valued their tools and took good care of them; many products of ancient toolmaking are objects of great beauty, still seen in museums around the world.

The blacksmith made the metal parts of tools, as well as agricultural implements, until factory forging became common in the 19th century. For many years after that, workmen still made their own wooden handles, and even today factory craftsmen use their machines to make some of their own specialized tools.

THE PLANE In the 1st century AD Roman craftsmen already possessed a range of carpenter's planes not far removed from modern ones: a wedged cutting iron with a cleared-away holding stock to allow the unrestricted escape of shavings. About 1860, in Boston, Massachusetts, Leonard Bailey filed the first of several patents for bench planes; his name can still be found on many modern factory-made planes as an acknowledgement of his contribution.

The modern bench plane has a steel cutting blade set into a triangular casting, called a frog, which in

turn is mounted in the plane base. The blade is pressed from nickel-chrome steel, surface ground to a precise thickness, and has a bevelled cutting edge, hardened and tempered. The edge must be honed by the user, the factory edge being unsuitable for fine work. The frog and the base are cast from grey iron and milled on the appropriate surfaces to provide accurate seating of the frog and to provide smoothness on the surfaces of the plane which come into contact with the wood.

Rosewood was traditionally used for handles and knobs on planes, but plastic mouldings are often used nowadays. The latest development in planes has been the introduction of disposable blades, which does away with the need for sharpening.

HAMMERS The hammer is a striking tool comprisprising a shaft, which acts as a lever-extension of the user's arm, and a hardened and tempered steel

Left: the modern bench plane is much the same as that used by the Romans. The plane base, top, is shown with the 'frog' and a selection of blades, which have to be honed by the user.

Below: automatic 'parting and pointing' of round bar in the making of Pozidriv screwdrivers. The tip of the bar is formed into a winged cross to fit the screw head.

Right: the head of a carpenter's claw hammer, forged from a bar of steel heated to a state of plasticity. The head is broken off and the next one forged from the same bar.

head. The shaft is usually made of straight-grained ash or hickory; these timbers have high cross-sectional strength with a degree of resilience to absorb shock. Hardened metal tubing and glass fibre reinforced plastics are also being used nowadays. Hammer heads are made by drop forging using bar steel; forging the steel while it is in a hot state of plasticity improves the molecular structure, causing the grain flow to redistribute itself for best mechanical properties. Heat treatment of the hammer face prevents it from breaking up or chipping when it is struck against other metal objects. The hammer head is usually fastened to the shaft by means of three wedges: a large wooden wedge and two smaller ones made of malleable metal, acting to spread the wood in the eye section of the head.

There are many types of hammers, but the most common are the carpenter's claw hammer and the pein hammer. The carpenter's hammer has a split claw opposite the face, specially designed to pull out nails; pein hammers (also spelt peen and pean) are used in the engineering trades. Pein is a word probably of Scandinavian origin which means to draw, bend or flatten metal or leather by beating it; the straight pein and the Warrington or cross pein hammers have wedge shaped blades opposite the striking face, while the ball pein is rounded. Other special purpose hammers for the engineering trades are made of rubber, plastic or soft, heavy alloys such as babbitt, which has a tin or lead base and is also used for bearings.

FILES Flat files for woodworking were made of copper in Egypt as long ago as 1500 BC. Modern files for removing stock from wooden or metal surfaces are pressed from strip steel; the specialized pressing operation results in a corrugated surface, and the high spots are sharpened by grinding and the blade is then hardened and tempered. Some files are produced with tangs on the end for wooden handles and others with holes in each end for fastening to wooden or diecast frames. The variety of files available is enormous; the cross section of the blade and the shape of the teeth are determined by the intended use of the tool. Mill or saw files are used to sharpen teeth on saw blades, for lathe work and smooth filing in general. Knife files have a cross section like a knife blade and are used by tool-and-die makers in corners and for cleaning worn screw thread. Bastard files, rasps and wood files have coarser teeth for rapid removal of stock. Curved tooth files come in several varieties; when moved across the work towards one edge they remove stock rapidly; in the other diagonal direction they leave a smoother finish. Special purpose files are made in many cross-sectional shapes, including square, triangular, half-round, round and rat-tail (round and tapered).

A modern development of the file is the shaping tool available in do-it-yourself shops with a very coarse surface, designed for use on plastic, glass fibre and other do-it-yourself materials. The Surform range of tools is the best-known example.

WRENCHES Wrenches are tools of various sizes for turning nuts and bolts. They are always forged of high quality steel for strength. One-piece wrenches are either open-ended or box end (enclosed), for more gripping power where there is room to fit the box shape over the nut. A combination wrench has one end open and the other end a box of the same size. The open end on a wrench is usually offset at an angle to the body, for more versatility when the nut or bolt is hard to reach. An adjustable wrench has a separate lower jaw which is adjusted by turning a vertical screw parallel to it. An adjustable wrench should always be used in such a way that the torque is applied by drawing the wrench toward the user, and with the lower (adjustable) jaw toward the user, so that most of the pressure is borne by the larger upper jaw; otherwise the wrench may slip off the nut, giving rise to its nick-name 'knuckle buster'. A pipe wrench is an adjustable wrench with hardened corrugations on the faces of the jaws; the jaws are set at a slight angle toward one another for gripping a pipe or a round or worn nut. In Britain, the wrench is often called a spanner, but a true spanner wrench is a special tool with a hinged upper

jaw having a peg on it which fits into a hole on a special nut.

An Allen wrench or Allen key is a case-hardened piece of hexagonal bar steel which fits into a hexagonal recess in the head of an Allen screw.

SAWS Modern saw blades are stamped out of high quality steel strip. Lower-priced saws are of uniform thickness and the teeth are set and sharpened on machines; the latest development has been Teflon-coated blades to reduce friction. Professional quality carpenter's saws are blanked out and clamped in a special vice; the teeth are set and sharpened by hand; the blade is supplied in a polished condition, and the owner keeps it free of corrosion by applying a thin film of grease or oil while it is not in use. The blade is usually step-tapered away from the toothed edge.

There are three main types of saws available for wood: the rip saw, the universal and the fleam. In all three types the teeth are set by bending one third of the length of each tooth in an alternate direction from the next tooth. In the rip saw, for sawing wood along the grain, the teeth are sharpened straight across the face, at a right angle to the axis of the blade. The fleam and the universal types have the teeth sharpened at a 60° angle to the axis of the blade. The type of file used for sharpening depends on the number of teeth per inch on the saw. Saws are classified according to a 'point' system; a ten point saw, for example, has nine teeth to the inch.

When sharpening the universal and fleam types, the face of one tooth and the rear of the next are sharpened, working down the length of the blade; then the blade is reversed in the vice and the rest of the teeth are sharpened the same way. The difference between the two types is that the fleam, a log-cutting type which is now becoming popular in the UK, has a smaller included angle of the tooth, resulting in in sharper teeth.

Hacksaws are used for cutting metal. The blades have holes in each end by means of which they are held under tension in a steel frame with a plastic handle. The American National Standard for hacksaw blades, for example, specifies that blades having 24 or 32 teeth per inch shall be set wavy, and other types shall be alternate set or raker set (every third tooth unset). A welded composite blade is made of two or more pieces of steel joined together, but the cutting edge or edges are made of high speed steel; a standard steel blade must contain not more than 1.25 tungsten or an equivalent alloy; and a high speed steel blade must retain its hardness at temperatures up to 1000°F (537°C). For general work where the blade is not changed for each job a blade of 18 teeth per inch is recommended.

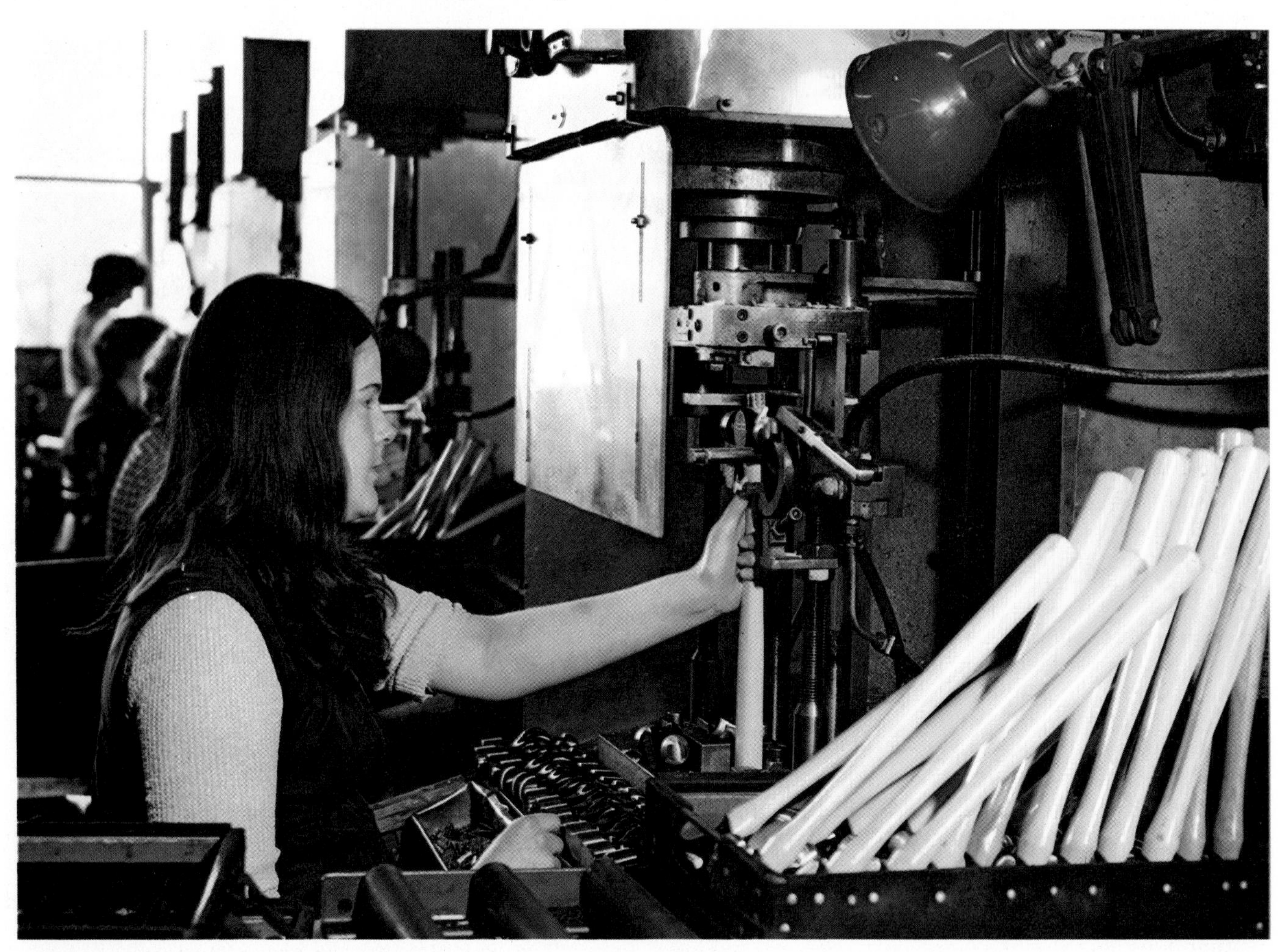

This page, near left: these heavy heads, for mauls or sledgehammers, are being heat treated. Most tools are heat treated in some way to give the metal a desired quality such as hardness, toughness or springiness.

Opposite page, far left: using a machine to fix the hammer head to the shaft. Three wedges are driven into the end of the shaft to cause it to spread, fixing it tightly in the 'eye' of the hammer head.

SCREWDRIVERS The screwdriver is a hardened and tempered steel blade made from bar stock. The working end is formed by forging or grinding and comprises a narrow rectangular section, made in various sizes to match the slots of screw heads. The handle end of the screwdriver has localized projections called wings formed on the bar; the traditional wooden ball handle is still widely fitted, but injection moulded plastic handles are more common.

Modern variations of the conventional screwdriver are the Phillips and Pozidriv designs; in these, a winged point has been substituted for the rectangular blade-tip. The winged point is formed by milling and driving the point into a sizing die, followed by heat treatment. It complements an equivalent female form in the screw head, giving more positive contact between screw and driver. These designs were developed to allow more torque for high speed assembly, especially using power-driven tools.

OTHER TOOLS Special purpose knives are made by tool companies, comprising a hardened and tempered blade of high carbon steel with a precision ground cutting edge. The blade is hardened in an electric furnace in an inert atmosphere to prevent surface contamination, and is simultaneously tempered; the cutting edge has the characteristics of a razor blade, but the blade is made of thicker material and is more durable. Blade handles are traditionally die-cast with provision for positive location of the blade; some are designed to allow full retraction of the blade when not in use, for safety and to prevent accidental damage to it.

The spirit level assists in achieving true vertical and horizontal planes during construction processes. Adjustable models can be set to check any prescribed angle. The essential component is the vial, a sealed plastic or glass tube containing a dyed liquid such as kerosene. The volume of liquid is controlled at the filling stage to allow an air bubble; the tube is formed in a radius to allow the bubble to take a position at the tangent of the radius. The vial is securely set and precisely located in a parallel extruded aluminium frame under known level conditions.

Chisels are made of steel and are forged, hardened and tempered; a bevelled cutting tip is provided, but like that of the blade in the bench plane, it must be honed by the user to his requirements.

A brace is a drilling device which comprises a steel body which is bent from bar stock into a crank shape, with a centre handle and a bearing assisted thrust pad on the opposite end from the drill for applying pressure. The crank throw varies from 6 to 14 inches (about

15 to 35.5 cm); the wider the throw the more torque can be applied. The hand drill has, instead of a crank, gears which are driven by a cranked handle; the gears are made of annealed cast iron. Both devices have self-centring chucks with sintered jaws held in place by springs; the hand drill also has a wooden or plastic handle projecting at a right angle from the body opposite the main gear for steadying the tool and applying pressure.

WINEMAKING

Wine is as old as civilization itself and has from the earliest time been one of the commonest subjects for literature and painting. We begin to get more tangible evidence with the expansion of the Greek Empire during the 1st millenium BC, and from the era of the Roman Empire we learn still more.

The Greeks themselves called Italy 'The Land of Vines' and it is true to this day that Italy grows vines from one end of the country to the other, indeed to such an extent that it has now surpassed even France in the quantity of its output. However it is France that has the greatest share of the world's most highly acclaimed wines; it was therefore an event of considerable significance when the Romans introduced the vine to the province of Gaul.

During the Dark Ages, it was the Church that was largely responsible for the survival of most of the skills of civilization, not least that of wine making, and many of the great vineyards in Europe owe their existence to the Church. But the evolution of wine as we know it today really began to take shape during the 17th and 18th centuries. After the discovery of the cork, it became apparent that wine could be aged in

Top: grafting a European vine on to an American root-stock. The two parts are cut to fit exactly by the machine in the foreground.

Near right: separating grapes from the stalk. This process also partly crushes the grape and allows the yeast on the skin to come into contact with the grape juice, thus starting fermentation.

Opposite page, far right: grapes are crushed and the stems removed, then either vatted or first pressed and then vatted. After fermentation, the 'cap' is removed and pressed to extract remaining juice. The wine is matured in barrels, fined (cleared) and bottled.

bottle for a long time, thus acquiring different characteristics from wine that had hitherto always been aged solely in cask or jar. Early bottles had had ineffective clay or wax seals. These had allowed some air to get at the wine, oxidizing the alcohol to acetic acid and thus turning it to vinegar—from the French *vin aigre*, 'sour wine'. (Commercial vinegar is made from wine or another alcoholic liquid by speeding up the reaction with acid-producing bacteria of the genus *Acetobacter* and a nematode worm, the 'vinegar eel', *Anguilula aceti*.)

The use of corks led to a change in the shape of the wine bottle, the shorter, squat bottle giving way to the longer, more cylindrical variety that was most suited to being laid on its side in a cellar, this being necessary in order to keep the wine in permanent contact with the cork to prevent it from drying out, and so shrinking and admitting air.

By the latter half of the last century the wine industry was booming, and it was at this point that the *phylloxera* disaster struck. A species of aphid suitably named *Phylloxera vastatrix* ('the devastator') coming from North America, succeeded in destroying virtually every vineyard in Europe by attacking the roots of the vines. Fortunately it was discovered that the native American vine had a root that was immune to the bug, and consequently most European vines are grafted onto an original American root-stock.

FROM GRAPE TO WINE Wine is the fermented juice from the grape of the *Vitis vinifera* species of vine, of which there are several thousand named varieties. Other beverages, often home-made, may borrow the word 'wine' for the sake of convenience, but this article is only concerned with wine in its true sense.

To make wine it is necessary to understand the two processes of viticulture, that is, the growing of the grapes, and vinification, or actual making of the wine. It is the skill employed in these two processes which contributes vitally to the ultimate quality of the wine

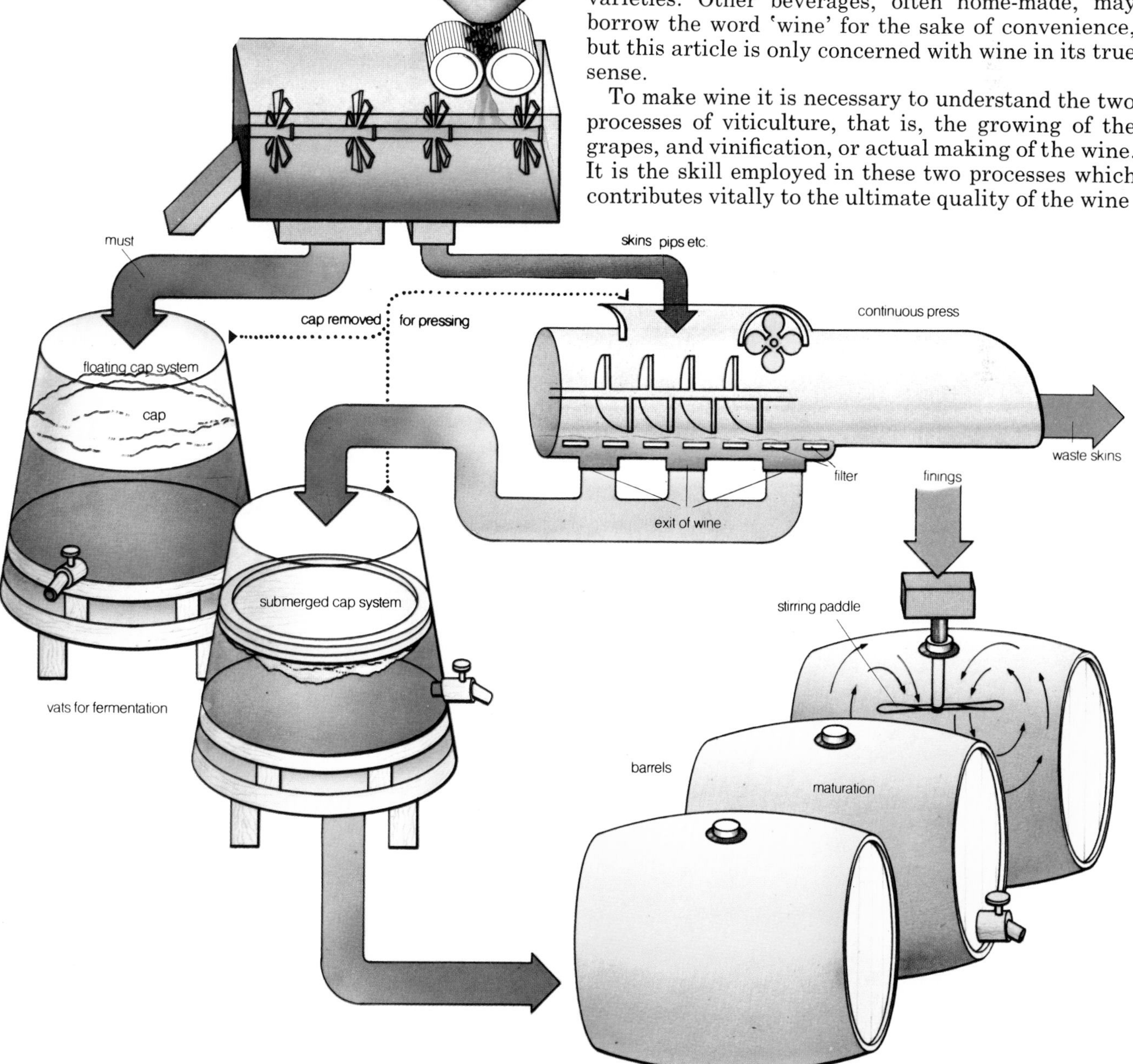

The other principal influences are the climate of the area concerned, the geological structure of the vineyard, and the variety or varieties of grapes grown there.

VITICULTURE Much care and control has to be exercised all the year round to have any hope of a successful harvest. Methods of pruning are of particular importance, and in the better areas vines are cut back very severely each winter, since a small yield of grapes per vine produces juice of significantly higher quality. The vines themselves remain dormant during the winter months.

During the first part of June (in the northern hemisphere) the flowering takes place and warm, calm

Right: the must (juice) being checked for sugar content during fermentation.

Below left: white wine is made in a press so that the skins are left behind.

Below right: laboratory testing of wine to measure acidity, specific gravity, etc.

weather is most important. The vines must be regularly sprayed throughout the summer against insects, mildew and 'oidium', a type of fungus which attacks young roots, leaves and grapes. By about 100 days after the end of the flowering, the grapes are usually ripe and the harvest can commence. This normally takes place therefore towards the end of September or early in October, but in certain areas it may be considerably later.

VINIFICATION Grape juice turns into wine by the natural and quite spontaneous action of fermentation, the process by which the sugar content of the grapes is converted by yeasts into alcohol and carbon dioxide gas. The yeasts come from the skin of the grape itself, although cultured yeasts may often be added, and fermentation will start as soon as the skin is pierced and the yeast has access to the pulp.

Fermentation will normally continue until the sugar has been converted into alcohol, or until the alcohol reaches a level of around 15% of the volume, at which point the yeasts are normally killed. It is, however, possible to arrest fermentation before all the sugar is used up, and thus make a sweet wine. This may be achieved by the addition of alcohol or sulphur, or by fine filtration to remove the yeasts.

The juice inside the grapes is normally more or less colourless, whether the grapes are black or white, and it is perfectly possible to make white wine from black grapes, as in the majority of Champagne, for instance. Nevertheless, red wine is normally produced from black grapes, and white wine from white grapes. The colour necessary for making red wine is obtained from the skins of the grapes, and hence the techniques for making red wine differ in most cases from those for making white.

To make red wine the grapes are normally fed through a machine which removes the stalks and slightly crushes the grapes. In France, this machine is known as a *fouloir-égrappoir*. The crushed grapes are then pumped into a vat where they ferment with their skins. As the juice begins to turn into wine, more and more colour becomes extracted from the skins. After sufficient colour and tannin (a naturally occurring acid which gives the wine a characteristic taste) have been acquired, the *must*, as the juice is called, is run off from the skins, which are then pressed to extract more wine, which will be deeply coloured and full of tannin. This wine is sometimes added to the other to provide more 'body', if so required.

The initial period of fermentation usually lasts about 10-14 days, after which the wine is usually racked and run off into wooden barrels. Racking is the process of separating the wine from the sediment which forms as a result of fermentation.

To make white wine, the grapes are also destalked and crushed, but they are then commonly put into a horizontal press where two end-plates gradually draw together and squeeze out the juice from the skins, which are left behind. Rosé (pink) wine is normally obtained from black grapes, allowing the juice to remain in contact with the skins for just as long as it is necessary for a sufficient amount of colour to be extracted.

Fermentation vats are commonly made of wood, stainless steel, glass fibre or concrete, sometimes glass-lined. It is most important that the right temperature is obtained, usually around 77°F (25°C), and most modern establishments have efficient methods of temperature control. Once fermentation is under way, the skins will float to the top of the vat, forming a cap. This prevents essential oxygen reaching the must underneath and it is important therefore to keep this cap broken up and the skins in contact with all the juice.

It is not normally necessary to add sugar to the must, but the practice is quite common particularly in northern vineyard areas where there is less sun and the grapes do not always ripen. The normal purpose of adding sugar is to increase the degree of alcohol. The process is commonly referred to as chaptalization, named after Jean André Chaptal (1756-1832). A secondary, or malolactic, fermentation normally takes place in the spring following the vintage, though it can sometimes be induced to take place immediately after the ordinary alcoholic fermentation. This malolactic fermentation is desirable to convert surplus malic acid, which would affect the quality of the wine, into lactic acid.

MATURATION The Bordeaux method, which is also followed in many areas, is for red wine to be drawn off into *barriques* or hogsheads in about the February following the vintage. These are wooden barrels, usually made of oak or chestnut, in which the wine matures by breathing in a small amount of oxygen through the pores of the wood. Evaporation

Below left: the wine is matured in wooden barrels which allow a small amount of oxygen to reach the wine through pores in the wood. The barrels are stored in cellars where the temperature is steady, generally for up to two years, but sometimes longer. The wine is fined to remove any cloudiness before being bottled.

Below right: cleaning the bottles in preparation for bottling. Red wines are further matured in the bottle; white wines are drunk young.

also takes place and it is necessary to keep the barrels regularly topped up. This period of maturation may vary: eighteen months to two years is quite normal but it may be the practice to age the wine for much longer, if so desired.

Before bottling the wine, it is normal to 'fine' it with an albuminous substance such as white of egg or gelatine. This should render the wine clear and bright. It is often desirable for wine to be matured still further in the bottle. The period of maturation for white wines is normally much shorter. In fact it is quite normal to bottle them in the spring following the vintage, if not even sooner, and they are often never matured in wood at all. Generally white wines are made to be drunk fairly young while they retain their freshness. Indeed many red wines are also vinified nowadays in a way that makes them attractive to drink while they are young.

With the best red wines, it always takes time for the compounds of acids, sugars, tannins, esters and aldehydes to resolve into an ultimately harmonious balance. Esters and aldehydes are by-products of the alcoholic fermentation which contribute particularly to the *bouquet*, or aroma, of a wine. If a wine is left for too long however, the colour will fade and the wine will begin to taste lifeless and insipid as the fruit and acidity lose their strength. When maturing or storing a wine it is important to have a place, ideally a cellar, where the temperature remains fairly constant and neither too cold nor too warm, preferably between 45°F and 65°F (7°C to 18°C).

FORTIFIED WINES In addition to ordinary wines, which have an average alcohol content of between 10 and 14%, there are also fortified wines with a volume of alcohol between 16 and 24%. Such wines as port, sherry and madeira are stronger than ordinary table wines since they are fortified by the addition of alcohol, usually grape brandy. This addition prevents further fermentation and thus leaves any residual sugar intact, so that the wine can remain sweet if so desired.

Port is made by mixing brandy with red wine which has only been half-fermented and thus retains half of its considerable sugar content. The grapes are grown in the valley of the Upper Douro river in Portugal, but the wine is nearly always taken down to the Port lodges near Oporto to be matured in pipes, 115 gallon (523 litre) barrels, where they may remain for between 2 and 50 years.

Vintage port is the product of a single year which was considered particularly good. This is bottled after only 2 years, but it is necessary to keep the wine in the bottle for a further 15 to 20 years before it reaches sufficient maturity. But there are other varieties of port which can be consumed sooner. If kept in wood longer, the wine will mature much faster and it can be bottled when it is ready to drink. Ruby port is comparatively dark and rough, being aged for a relatively brief time in wood, but tawny port is kept for much longer, thus achieving a much paler colour.

Sherry, from the Jerez district of Spain, is also a fortified wine, though some dry *fino* sherry need not be fortified as a high enough degree of alcohol may be achieved without it. The most distinctive feature of sherry is the so called *solera* system whereby the wine is kept in butts, the older ones being continually topped up from younger ones of the same style. Thus, by continuous careful blending, the same sort of wines can be produced consistently year after year. Sherry is produced in varying degrees of dryness or sweetness according to demand.

Madeira, from the volcanic island of the same name, is also commonly produced by a solera system, though occasionally wines of a single vintage can be found. They are often none the worse even after 100 years or more. The wine is subjected to a prolonged period of up to 4 or 5 months in *estufas*, or stores, where the temperature is sustained at around 120°F (49°C). This helps give the wine its characteristic flavour of slightly burnt caramel.

SPARKLING WINES Champagne, and other sparkling wines made by the *méthode champenoise*, as it is called, are given a second fermentation in the bottle. The carbon dioxide gas produced is unable to escape and dissolves in the wine, thus producing the familiar sparkle. The second fermentation is carried out with the bottles upside down, so that the sediment is deposited on the cork. The neck of the bottle is then frozen and the cork drawn, along with the frozen sediment. A new cork is inserted and wired down; the wire and the special heavy bottle are required to resist the gas pressure.

Some other sparkling wines are made by a system known as the 'cuve close' method. The principle is the same, but it takes place in a vat, or cuve, rather than in the actual bottle. This is clearly a more economical method of production, but the bubbles are likely to be less prolonged once the bottle is opened. A still cheaper method of producing sparkling wines is simply to inject carbon dioxide into the still wine, but the result is much less satisfactory.

MODERN REQUIREMENTS The methods described above are typical of those which are standard for the production of what can be called fine wine. But winemaking is not an art that has stood still over the years, and its evolution continues daily as new techniques are introduced and as the requirements of the market change. It must be remembered that the bulk of wine produced in the world is made for current consumption. For commercial reasons it is desirable for turnover to be as rapid as possible and accordingly the treatment that is given to the finest wines will not be given to the cheaper ones. It is thus quite normal to find practices such as filtration and even pasteurization common with ordinary wines that are not intended for further years of maturation in bottles. The frequent addition of sulphur as a preservative against oxidation is normal, particularly with white wines. But the less artificial assistance that a wine receives and the more natural its evolution the better. Good wine needs to be treated with care at all stages of its production and development.

WIRE and wire products

Wire is an important industrial product which is taken for granted nowadays, but it was only in the mid-nineteenth century, after the invention of the steam engine, that mass production of wire began. The incentive was provided after about 1840 by the demand for wire for cable (wire rope), for the telegraph and for wire fencing. After the American Civil War (1860-65) there were more incentives, such as the development of barbed wire for fencing, the insatiable demand for wire products such as pins and nails, and finally the invention of the telephone.

In ancient times wire was made by cutting strips from thin pieces of metal. For centuries the technique of drawing metals has been in use, but only short lengths of wire of small diameter could be made, because the process depended on the strength of a man pulling the wire through the die.

WIRE DRAWING Today wire is drawn on automatic machinery from hot-rolled lengths of metal called rods. Since the rods are rolled at high temperatures (2000°F; 1090°C), a hard scale forms on them which must be removed before they can be passed through the drawing dies; this is usually done by pickling (immersion in sulphuric acid). Grit blasting is also used, especially for spring wire. Then the rods are washed and coated with a solution chosen according to the type of steel and its intended use; if some types of solution are used, the rods may be baked in an oven. The coating may serve the function of carrier for the lubricant during drawing.

Long lengths of rods are made by welding them together end to end. The end of the rod is pointed, and the point is threaded through the die and attached to the drawing block. The die is usually made of tungsten carbide; dies made of diamond are sometimes used for small gauges of wire. The hole in the die through which the wire is drawn comprises four sections: a taper which allows the rod to pull some lubricant in with it, a second taper which actually performs the size

Below left: wire drawing machinery. The final drum, on the right, is spinning too fast for the coils to be seen. The dies are made of tungsten carbide.

Near right: a cable laying-up machine. Insulated power cable is pulled from drums mounted on the large foreground wheel as it turns, and the cable is finished in the background.

Below, far right: this wire extrusion machine can reduce a rod to a wire 150 times smaller at speeds up to 30m/sec (100ft/sec) for aluminium.

reduction, a short cylindrical section called the bearing which determines the size and roundness of the wire, and the exit or reverse taper. The exact configuration of the die is determined by several factors, such as the type of metal being drawn, the amount of reduction of size taking place, the kind of lubricant used and so forth. The exit taper is designed to provide strength for the die and to minimize wear, at the end of the bearing section. When the die does begin to wear, it can be enlarged and used in an earlier stage of drawing, or for larger diameter wire.

Extra strength is imparted to the die by shrinking a steel jacket around the carbide nib. To assemble a shrinkage fit, the dimensions of the parts are carefully calculated, the jacket is heated to an appropriate temperature and the assembly is done while it is hot. When the parts reach room temperature they will be fitted very tightly together. This technique is also used for fitting bearing surfaces to shafts, wheels or pulleys on hubs and so forth.

Wire can never be drawn without lubrication, because this would result in excessive wear to the die, shrinkage or breaking of the wire and other problems. Depending on the type of drawing being done, wet or dry lubricants are used; wet lubricants include soluble oils and soap solutions; dry lubricants include greases and soap powders. In addition, the die itself is water cooled. The draw block revolves, driven by an electric motor, pulling the wire through the die, reducing its size and increasing its length.

Even with lubrication and cooling, the wire may be hardened and stiffened by the drawing process. Depending on the ductility and the tensile strength of the metal being drawn, as well as the final size required, several drafts (passes through dies) may be necessary. Since each draft hardens the wire, it is softened by heating it to a point just below its critical range. This is called process annealing and may be done several times. Various other types of final heat treatment are also used, depending on the intended use of the wire being drawn.

Modern wire drawing machinery is designed with multiple blocks pulling the wire through several drafts. Since the length of wire is increased each time, for continuous production the speed of each block must be faster than the one before it.

NEEDLES AND PINS Many consumer products are made of wire. For example, in the manufacture of pins, hard wire is drawn into the machine, straightened by passage between rollers, and gripped by a pair of jaws. A pin-length is struck on the end by a header die to form the head; then it is cut off and the cut-off end passes files or cutters which form the point.

Iron needles were first made around 100 BC; needle manufacture has been carried on in the Redditch district of England for 300 years. A large factory at Studley, opened in 1950, makes needles beginning with drawing the wire from Sheffield steel rods 0.212 inch (5.36 mm) in diameter. From coils the wire is straightened and cut to the length of two needles. Each length is sharpened at both ends; pairs of matching dies stamp the eye impressions in the centre, leaving a tiny bit of waste material between the eye ends; then

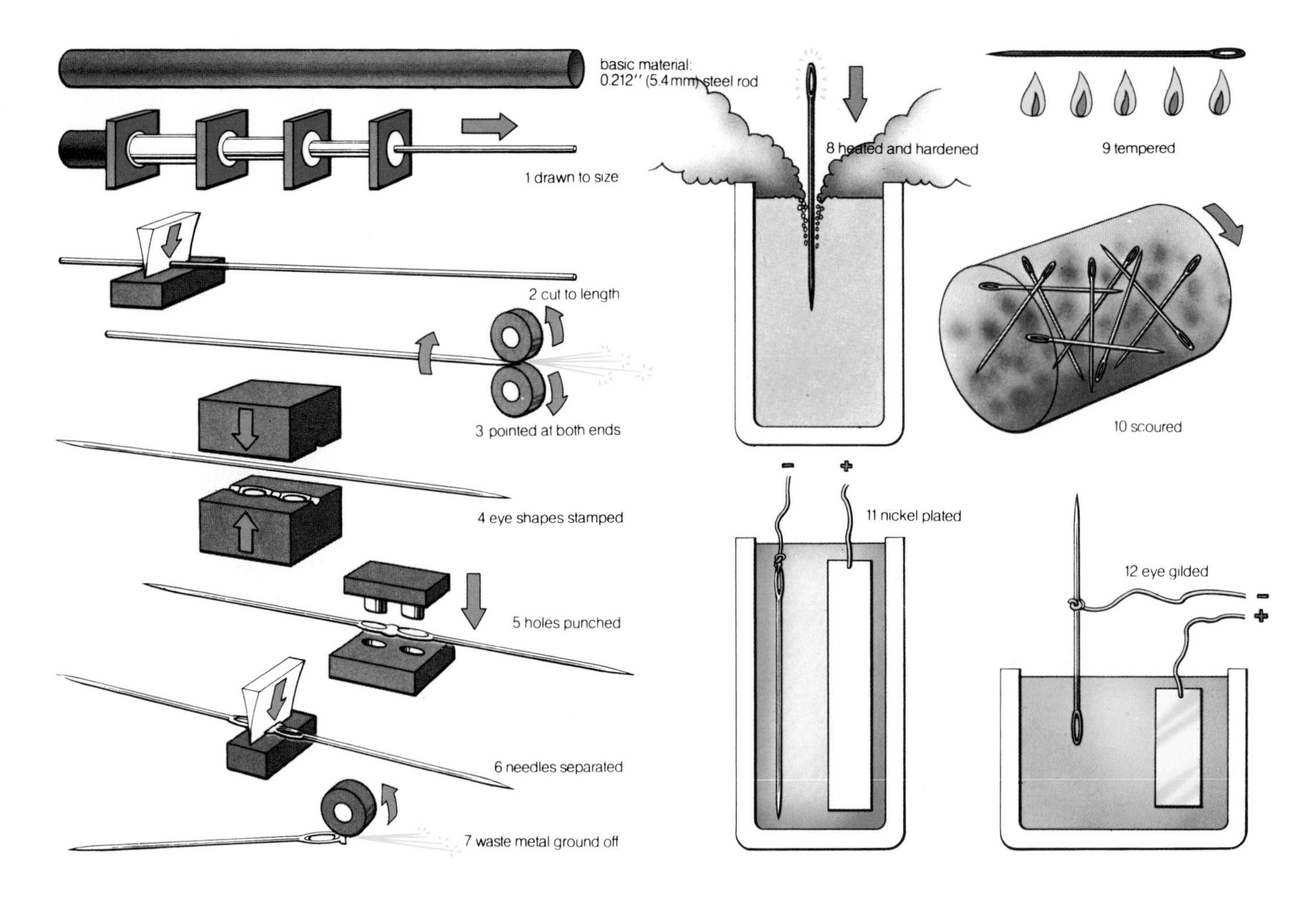

the holes in the eyes are punched. The wires are then broken into two separate needles and the waste is ground off around the eyes. The needles are hardened and tempered by heat treatment, and 'scoured', cleaning them as well as leaving them completely smooth and highly polished. All the needles are nickel plated and a high proportion of them are gilded. The factory exports millions of needles each week.

The machinery which makes pins and needles has not changed fundamentally since the nineteenth century, but has been highly developed until great production speeds are possible. Each manufacturer has his own techniques for fast production to keep costs down, and some of the machinery even does the packaging.

NAILS Nail manufacture is one of the most efficient and highly automated of industrial processes. Originally nails were forged by hand; then they were cut from sheets of rolled iron. Mass production came in the nineteenth century, when a method of producing nails on automatic machines from coils of wire was developed. The basic principle of the nail forming machine has not changed since then.

Wire is drawn from a coil into the machine. The head is formed by flattening the end of the wire against a die; then the wire is pinched and cut to the correct length by ground cutters moving together simultaneously to form the point. Finally, the expelling mechanism knocks the finished nail into a pan at the bottom of the machine. The basic operation is augmented by additional forming or twisting devices for special nails, such as roofing nails with twisted shanks and those with helical spirals to give an extra-firm grip.

After forming, nails can go through a variety of treatments, depending on the intended use. They are cleaned in a rotating barrel of hot caustic soda to remove grease and wire 'nippings'. Then they are often tumbled in a second revolving drum containing heated sawdust to give them a bright finish. Other finishes often applied to nails include galvanizing, cement-coating and sherardizing, a process of applying a corrosion-resistant coating of zinc by heating the object to a temperature of about 300°C (570°F) in a closed container with a powder of zinc dust and zinc oxide.

The largest nail factory in Britain is at Cardiff; it contains more than 200 fully automatic nail forming machines, from those weighing only a few hundred pounds to giant machines weighing seven tons. The total production of the machines is about 1400 nails a second, or about 300 million in a normal week. This is about 800 tons by weight, or about half the total production of the country.

The packaging machinery in a nail factory is also highly automated. It includes magnetic elevators

Left: simplified step-by-step diagram of needle manufacture. The original steel rod is drawn, cut, pointed, shaped, smoothed, plated and gilded; even allowing for the fact that needles are made in large batches, it is remarkable how inexpensive this precision-made product is.

Right: nail manufacture. Wire is drawn from a reel into the feed box and the head formed by a die. The finished form, pointed by being cut off, is then expelled, washed in a revolving drum of caustic soda and polished in warm sawdust.

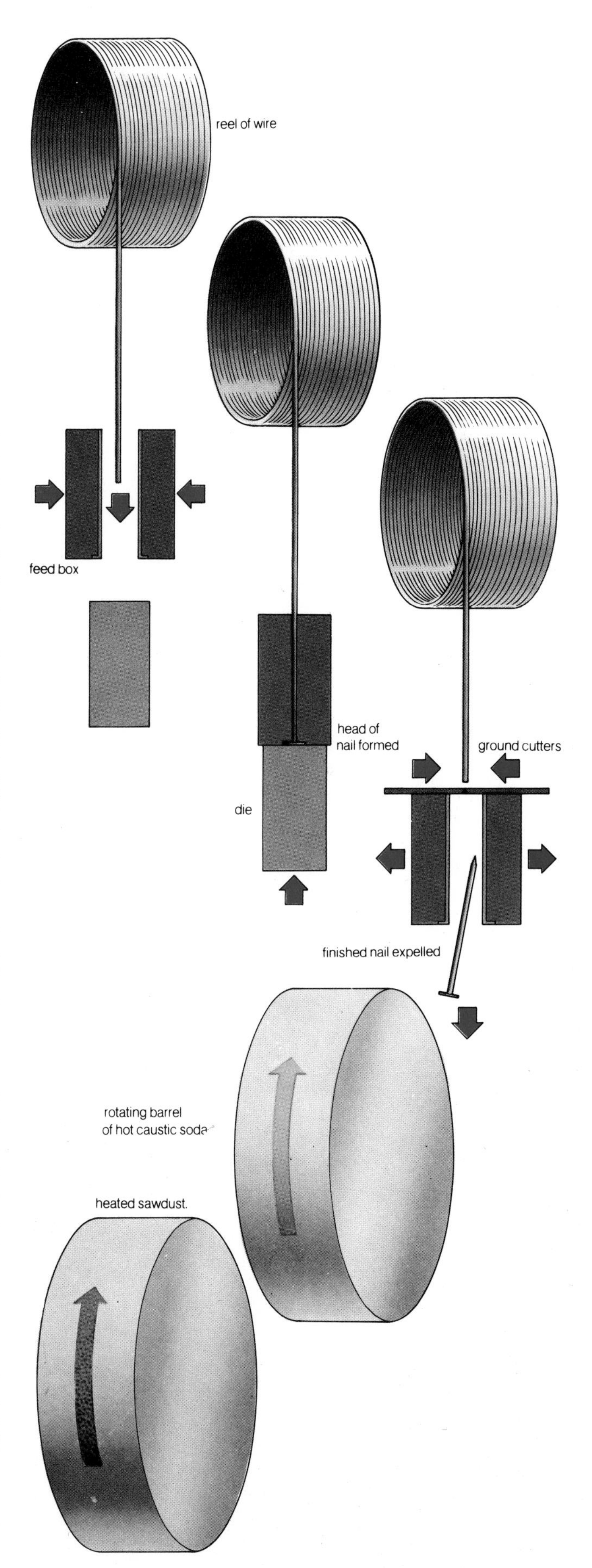

which convey the nails to weighing machines, where they are dropped into cartons past magnetic poles which automatically stack them in parallel rows. After packing the nails are demagnetized.

SPRINGS Coil springs are wound around a mandrel or former. For making small numbers of special-purpose springs, a lathe may be used. The mandrel is inserted in the chuck, one jaw of the chuck is loosened, the bent end of the spring wire is inserted and the jaw tightened again. The lead screw of the lathe is engaged according to the number of coils per inch of the finished spring, and the tool post of the lathe is used to guide the wire on to the mandrel.

For mass production of springs, special-purpose machinery has been developed in which the spring wire is pulled off a reel by means of rollers and fed into the machine. It passes over a stationary mandrel and strikes a deflector plate, which makes it curl itself around the mandrel. At a pre-determined point in the machine cycle, the wire feed stops to allow the end of the spring to be cut off. Attachments can form the ends of the wire into hook shapes, or bend them or grind them so they are square (at a 90° angle to the length of the spring).

Springs can be machine produced at a rate of several hundred a minute. Spring wire of up to about 16 mm diameter is coiled cold, but above that size the coiling is usually undertaken at temperatures equivalent to hot forging or strip rolling operations.

WIRE ROPE In the manufacture of wire rope, a number of wires are twisted helically to make a strand, and several strands are then twisted together to make a rope, in a fashion similar to making ordinary hemp rope. Wire rope has tremendous strength for its size and weight; it is used to connect control surfaces of large aircraft to their hydraulic actuators, for example, as well as for haulage and hoisting applications in cranes, cable railways, well digging, excavating and so forth.

Wire to be used in wire rope manufacture is subjected to heat treatment consisting of raising the temperature of the steel above its critical range and then cooling it rapidly. This produces a fine grain structure with iron carbide distributed through a ferrite matrix in a finely divided form known metallurgically as sorbite. The wire may be galvanized if required.

The machines used to make the strands and the finished ropes range in size from small enough to fit on a long bench to as large as locomotives. Spools or

Right: forming machine for making conical spiral springs such as those for upholstered seats. As the spring is formed the forming mandrel moves away to make a progressively wider spiral.

Below: wire rope being made by twisting the individual strands together on a machine.

Opposite page, below right: illustrations from Chester Carlson's original patent for xerography. Fig. 1: charging plate by rubbing. Fig. 2: exposing the plate. Fig. 3 and 4: developing the image. Fig. 5: transferring it to sheet. Figs. 6 and 7: fixing.

bobbins of wire or strand are installed, one in each section of the machine. The wire is threaded through holes the whole length of the machine; in operation, the outside of the machine spins, and wire is drawn off the bobbins and twisted together at the front end, where the wire rope or strand, constantly lubricated, is pulled through a die to consolidate it.

There are two common ways to twist or lay wire rope: the regular lay and the Albert or Lang lay. In the regular lay the helical twist of the strands is opposite in direction to that of the twist in the wires of the strand; in the Lang lay, both twists are in the same direction. Lang lay rope is more flexible and wears better on the pulley or sheave (a grooved wheel in a pulley).

There are four standard categories of wire rope, referred to by two numbers which describe their construction: the first number is the number of strands in a rope; the second is the number of wires in each strand. 6×7 rope is called a standard course laid rope, and is used for heavy hauling duties, such as well drilling and in tramways; 6×19 is used more generally than all others for hoisting; 6×37 is an extra flexible hoisting rope for use on small sheaves; and 8×19 is interchangeable with 6×37 except that it distorts from its construction more readily and therefore conditions of service which cause crushing of the rope require the 6 strand design. Additionally, wire rope is classed according to the carbon content of the wire from which it is made, in a range from 0.05 to 0.9%.

As an example of the load-carrying capacity of high-carbon steel wire rope, the breaking strength of the 6×37 design is given for two extremes of diameter: $\frac{1}{4}$ inch (6.3 cm) wire rope will lift 5500 pounds; 3 inch (7.62 cm) rope will lift 720,000 pounds. The working stress of a rope, however, should be restricted to one-fifth of the breaking weight; in other words, there should be a safety factor of five, and higher factors are required for some applications.

The core of a wire rope is greased and the strands are lubricated during manufacture, but the treatment will not last the service life of the rope. It must be kept clean to avoid corrosion, by wiping it with waste (rags) or hessian, but solvents such as kerosene should not be used. The rope should then be regreased with an acid-free lubricant such as petroleum jelly, applied by hand or wiped on with waste, and well worked into the spaces between the strands.

Failure of wire rope is caused by tensile stress if the rope is overloaded, by high compressive stress as the wires press in on each other, as in running in a V-groove, and by the wires rubbing against each other when bending around a sheave. Kinks and small loops should be avoided; increasing a pulley diameter by twice the circumference of the rope will double its life, while a rope subject to reverse bends (around one pulley in one direction, then around another pulley in the opposite direction) will have its life halved. If a rope requires cutting, at least three soft-wire fastenings should be placed around it on each side of the intended cut, to prevent the strands from unlaying.

XEROGRAPHY

During the early 30s the only method of copying business documents was basically photographic. This meant that the process was slow and used liquid processing chemicals.

A patent attorney named Chester Carlson, living in Astoria, a suburb of New York, realized the problems of these methods and spent a number of years developing his process. He called it electro-photography. His early experiments used a sulphur coated plate which he charged with static electricity by rubbing with a piece of fur. He then exposed it to a reflected image and the light destroyed the static charge. Lycopodium powder was blown over the plate and adhered to the charged areas. This image was then transferred onto a piece of waxed paper to produce a copy.

This very crude process was to be the basis for a multi-million pound company but he spent six years trying to find a financial backer. He was turned down by both IBM and Kodak but eventually found support with the Batelle Research Institute and a small photographic company called Haloid. The newly formed company called its process xerography from the Greek *xeros* and *graphe*, meaning 'dry writing'.

THE XEROGRAPHIC DRUM The modern process

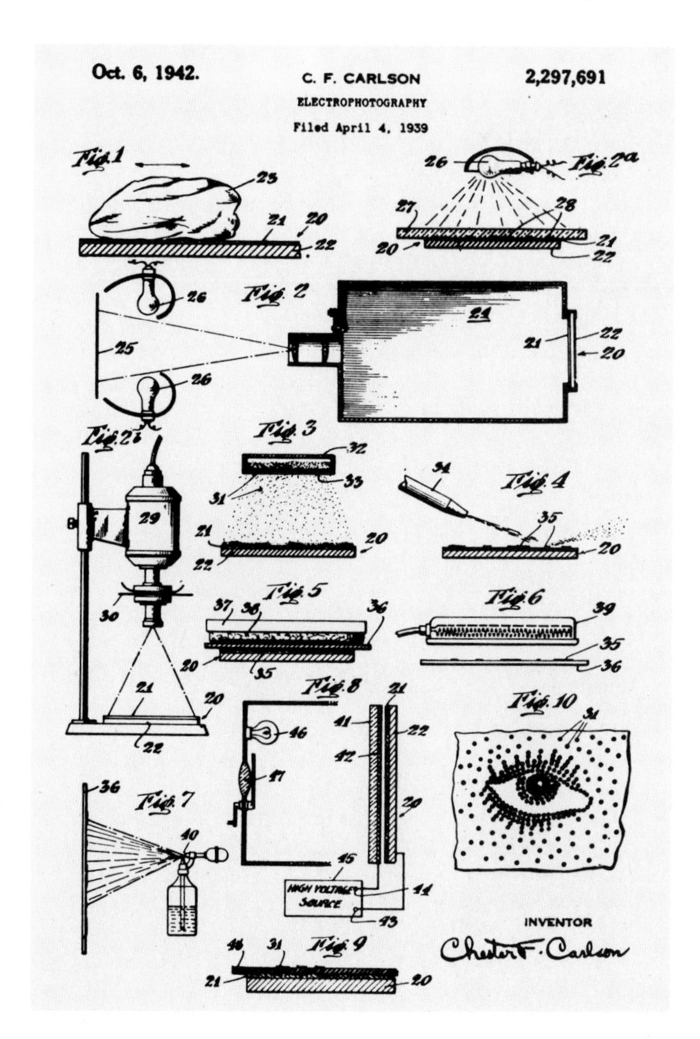

is very similar to the original but has been refined to six stages. Before discussing these it is necessary to look at the xerographic drum—the 'heart' of the machine.

The drum surface consists of a base of aluminium on which is laid a thin layer (approximately 10 atoms thick) of aluminium oxide. On top of this is a layer of selenium. Selenium is a photoconductor, that is, it will only conduct when exposed to light, and the drum coating is made especially sensitive. Thus the drum can have an electrostatic charge placed on the surface that will remain there—provided it is kept in the dark. If exposed to light, the selenium conducts the charge away to the aluminium, where it is neutralized. The aluminium oxide is an insulating layer to slow down the rate of discharge.

The first step of the xerographic process is to charge the whole surface of the drum electrostatically in the dark. This is done by rotating the drum under a corotron, a bare wire to which a high (approximately 7 kV) positive voltage is applied. This voltage ionizes the air, and a blue cloud or corona is often seen around the wire. The charge must be uniform on the drum in order to produce a uniform copy and this means that corotron must pass the drum at constant speed, the space between corotron and drum must be constant and the corotron voltage must be constant. Furthermore, there must be a good earth all over the aluminium and the corotron wire must be clean. This is a critical phase of the operation.

EXPOSURE AND DEVELOPING The next step is exposure. The image of the original to be copied is projected on to the drum by a series of lenses and mirrors. Like a photographic image it is reversed and inverted. The white areas of the original reflect a lot of light which destroys the charge on the drum, but black areas do not reflect and so leave the charge intact. There is thus an image in static electricity on the drum; it is a latent image in that it cannot be seen and would be destroyed if totally exposed to light.

During the next step, developing, a special dry developer is used. This developer is a mixture of a carrier and toner powder. The carrier consists of a mass of tiny glass beads (sometimes sand or metal shot) coated with plastic, about 0.25 mm (0.0077 inch) diameter. The toner is a fine black powder composed of a thermoplastic resin and carbon. The toner particles are extremely small—in relation to the carrier they are like a pea to a football.

Both the toner and the plastic covered beads are tribo-electric, that is they generate static electricity when rubbed together. The carrier receives a positive and the toner a negative charge. Thus the carrier beads become covered in a layer of toner.

The developer is poured over the drum and the toner covered carrier beads roll over the surface. The positive charge of the latent image is greater than that on the carrier so the toner adheres to the charged areas of the drum. (This corresponds to the black areas of the original.) Thus there is now a real image in toner on the drum.

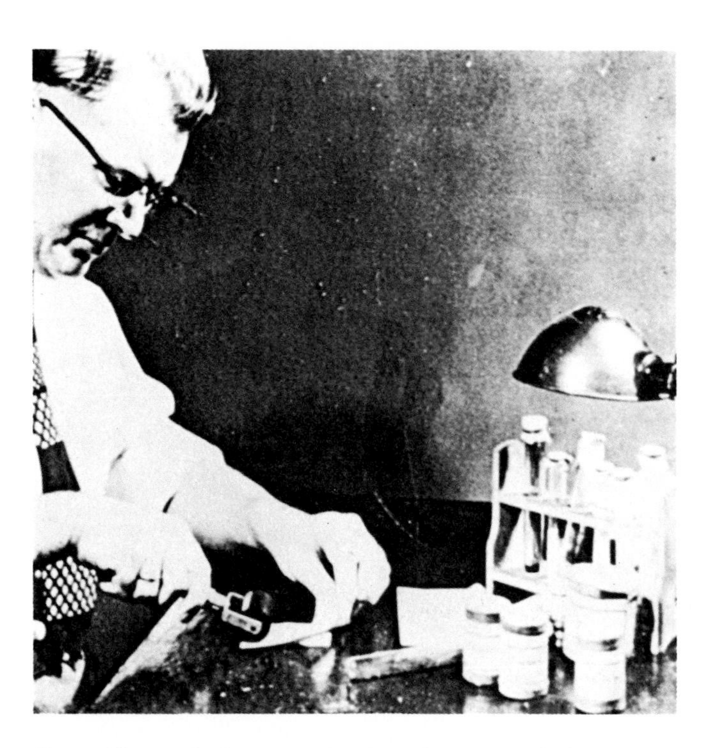

Above: Chester Carlson at work in his laboratory in the late 1930s. He is shown pressing paper with a roller against a plate which has been charged, exposed and developed. This was how he transferred the image originally.

Below: his first successful result.

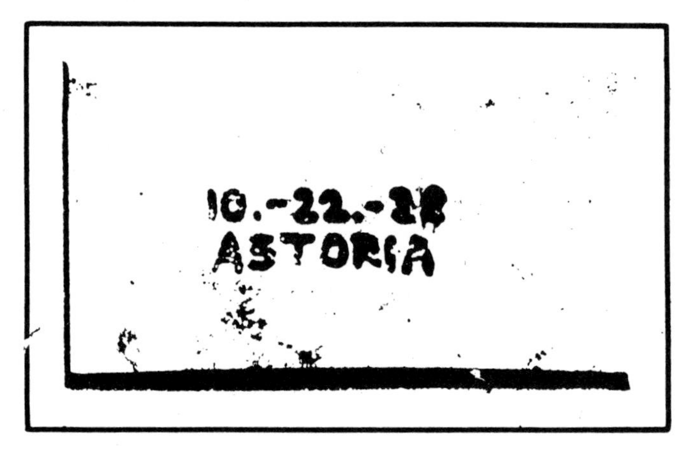

TRANSFER AND FUSION The next step is transferring. The toner is held on the drum by the positive charge. Thus to remove it and get it on to paper the paper must have a higher positive charge. This is done using a corotron again. It charges the paper as it is pressed against the drum and the toner now clings to the paper.

The copy is almost finished. It now passes through the fifth step: fusing. The toner is a thermoplastic resin, that is, it melts with heat. The copy is therefore passed under a radiant heater or through heated pressure rollers. The result is the toner melts into the fibres of the paper to give, when cool, a permanent dry copy.

CLEANING The copy is now finished but, because the transfer of toner is never achieved completely, the drum must be cleaned before the next copy can be made. Cleaning, the sixth and last step, is achieved in three stages. Firstly, the drum is discharged by means of a negative or AC corotron; this is followed by wiping and exposure to light.

The negative corotron destroys the positive residual charge and makes it easier for the toner to be brushed or wiped off the drum. Wiping can itself tend to induce a static charge, so exposure to light neutralises any charge left on the drum. All the six steps occur each time a copy is made, and some machines produce up to two copies per second.

XERORADIOGRAPHY An interesting development of xerography is xeroradiography. Here X-rays are used instead of visible light to provide a hard copy X-ray instead of the usual transparency. The process is the same but a flat plate is used instead of a drum. The plate is charged and held ready for use with others in a unit called the conditioner. When a X-ray is required the plate is taken out of the conditioner and placed under the patient. X-rays are directed at the plate and after exposure it is placed in the developer unit. Because of the need for detail, a special blue toner is used without a carrier. Cloud development is used, that is, a measured quantity is blown across the plate and clings to the charged areas. The image is transferred to a high quality glazed paper to produce the final image.

Its main advantages over the old process are that it requires lower X-ray doses, speeds are higher (a copy can be ready in a few minutes) and it gives a better quality image. Also, it is a daylight process (no dark-room required) and it provides a hard copy that can be selected as a negative or positive. It is particularly useful in soft tissue work, such as in the throat or larynx or mammography (breast X-raying), where conventional X-rays do not give good definition.

USING COLOUR ORIGINALS So far it has been assumed that black and white originals have been used in the copying process. When, however, producing black and white copies from colour originals there are complications because of the colour sensitivity of the selenium drum. Blue light causes rapid discharge of the selenium while red, at the other end of the spectrum (longer wavelength), causes little discharge.

This means that blue print is difficult to copy because it has the same action on the selenium as the white background on which it is printed. Red, on the other hand, behaves like black and copies well.

COLOUR COPYING Colour xerography has recently been developed which will provide copies with a range of six colours and black. The technique is similar to that used in colour printing and involves

Below: xeroradiography is xerography with X-rays, allowing lower dossages of radiation and faster results. These prints show tumours in breast tissue.

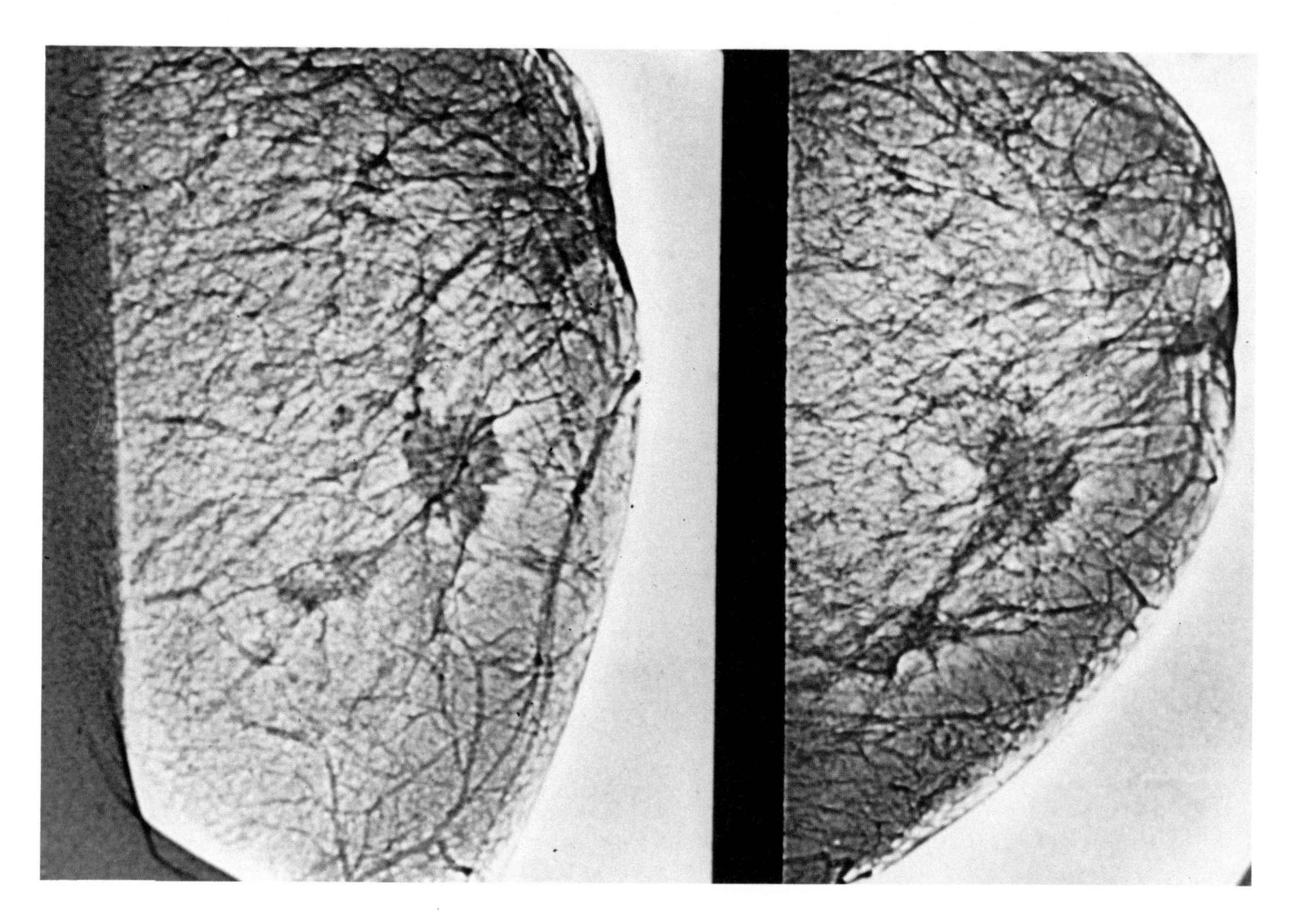

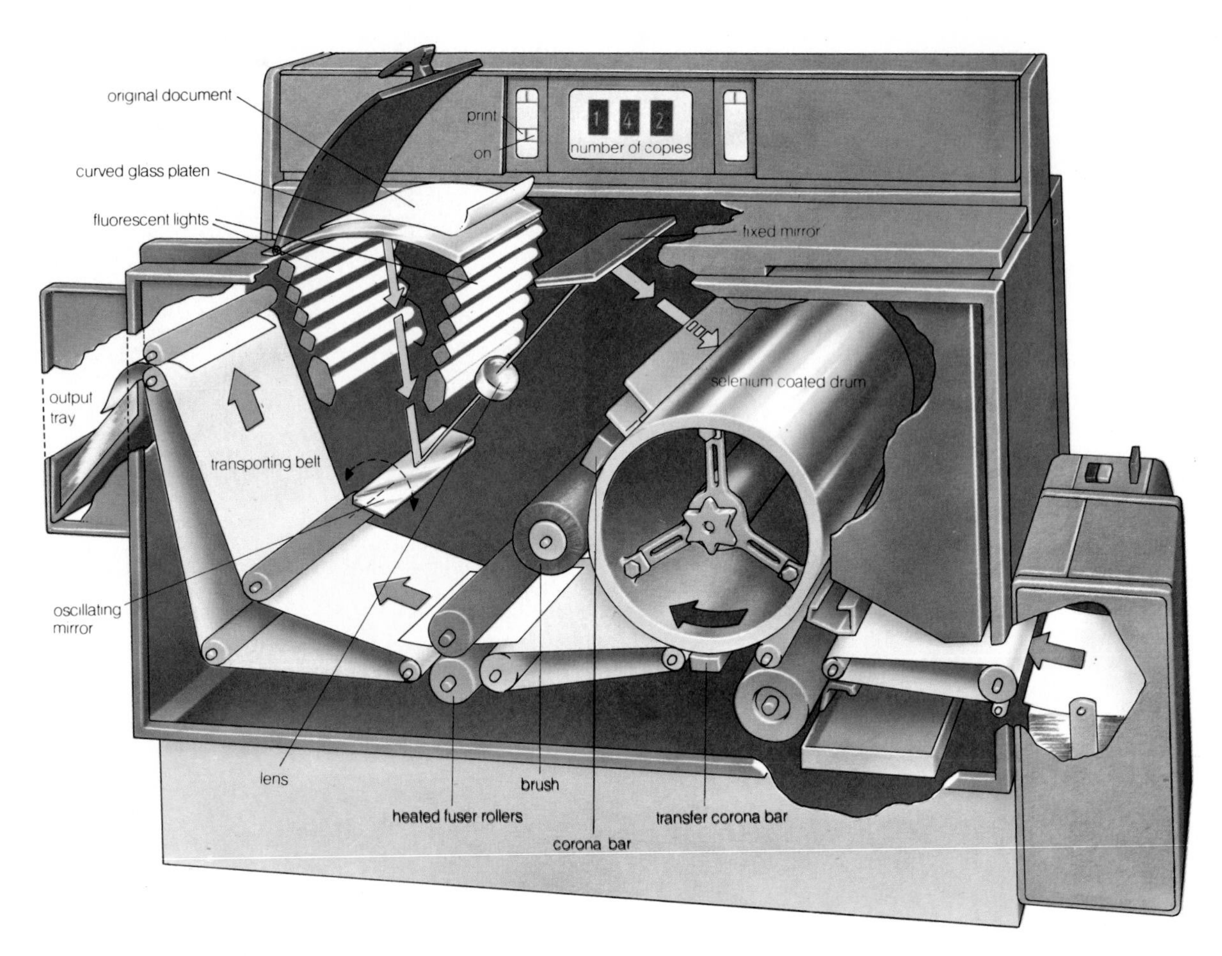

Above: a typical office xerographic copier. The document's image is scanned by the oscillating mirror and transferred via the lens to the drum, which has just been charged by a corotron. The charge image is then 'developed' (coated with toner) and transferred to the paper, which has been charged by the transfer corotron. Heated rollers then fix the image, while the drum is cleaned for re-use.

Below: a modern compact copier mounted on castors.

separating the original into three colour images. The colours are magenta, yellow and cyan—the primary colours of the subtractive colour system. Red, blue and green can be produced by overprinting two colours: magenta and yellow for red; cyan and yellow for green, and cyan and magenta for a darker blue than cyan alone. Over-printing all three colours gives black.

To obtain the colour magenta, for example, all areas on the drum must be discharged except those corresponding to magenta in the original. A magenta toner will adhere to these areas charged. Because magenta is a combination of red and blue a green filter is used. This will only allow green light to reach the drum and consequently the drum will be discharged at all areas corresponding to where green is present in the original (including white, which contains green). As green lies in the middle of the spectrum the charged areas correspond to the remaining parts of the spectrum—red and blue, which is magenta.

The other two primary colours necessary for colour printing are produced in a similar fashion. The yellow colour is produced from a yellow toner after exposing the drum through a blue filter. For cyan a red filter is used.

These three colours are added one after the other, exposing the drum through each filter in turn and developing at each stage with the correct toner. The composite is then fused and delivered as a full colour copy in about 30 seconds.

INDEX